The Asian Monsoon: Causes, History and Effects

The Asian monsoon is one of the most dramatic climatic phenomena on Earth today, with far reaching environmental and societal effects. But why does the monsoon exist? What are its driving factors? How does it influence the climate and geology of Asia? How has it evolved over long periods of geologic time?

Almost two-thirds of humanity lives within regions influenced by the monsoon. With the emerging economies of China, Vietnam and India now adding to those of Japan, South Korea and Taiwan, the importance of the region to the global economy has never been greater. Monsoon strength and variability have been and will continue to be crucial to the past and future prosperity of the region.

The Asian Monsoon describes the evolution of the monsoon on short and long timescales, presenting and evaluating models that propose a connection between the tectonic evolution of the solid Earth and monsoon intensity. The authors explain how the monsoon has been linked to orbital processes and thus to other parts of the global climate system, especially Northern Hemispheric Glaciation. Finally, they summarize what is known of the monsoon evolution since the last ice age and note how this has impacted human societies, as well as commenting on the potential impact of future climate change.

This book presents a multi-disciplinary overview of the monsoon for advanced students and researchers in atmospheric science, climatology, oceanography, geophysics and geomorphology.

PETER D. CLIFT is Kilgour Professor in the School of Geosciences at the University of Aberdeen, a Research Affiliate at the Massachusetts Institute of Technology and a visiting professor at the Chinese Academy of Sciences, Guangzhou. His research focuses on the integration of marine and terrestrial data sets to understand how the monsoon changes with time. He has authored over 100 peer reviewed papers and has acted as lead editor on two other books, as well as a special collection concerning the monsoon for *Palaeogeography, Palaeoclimatology, Palaeoecology*. He is co-leader of IGCP 476 "Monsoons and Tectonics".

R. ALAN PLUMB is Professor of Meteorology and Director of the Program in Oceans, Atmospheres, and Climate at the Massachusetts Institute of Technology. He has been an editor of *Journal of the Atmospheric Sciences* and of *Pure and Applied Geophysics* and has published approximately 100 peer reviewed papers, as well as co-editing one previous book and co-authoring an undergraduate textbook on *The Circulation of the Atmosphere and Ocean* (Marshall and Plumb, 2008).

The Asian
Monsoon

Causes, History and Effects

PETER D. CLIFT
University of Aberdeen, UK

R. ALAN PLUMB
Massachusetts Institute of Technology, USA

CAMBRIDGE
UNIVERSITY PRESS

32 Avenue of the Americas, New York NY 10013-2473, USA

Cambridge University Press is part of the University of Cambridge.

It furthers the University's mission by disseminating knowledge in the pursuit of education, learning and research at the highest international levels of excellence.

www.cambridge.org
Information on this title: www.cambridge.org/9781107630192

First published 2008
First paperback edition 2014

A catalogue record for this publication is available from the British Library

Library of Congress Cataloguing in Publication data

Clift, P. D. (Peter D.)
The Asian monsoon : causes, history and effects / Peter D. Clift and R. Alan Plumb.
    p. cm.
   Includes bibliographical references and index.
   ISBN 978-0-521-84799-5 (hardback)
1.   Monsoons–Asia.   I. Plumb, R. Alan, 1948– II. Title.

   QC939.M7C55 2008
   551.51′84–dc22

                              2008000807

ISBN 978-0-521-84799-5 Hardback
ISBN 978-1-107-63019-2 Paperback

Contents

Preface

The Asian monsoon is one of the most dramatic climatic phenomena on Earth today, with far-reaching environmental and societal effects. Almost two thirds of humanity live within regions influenced by the monsoon. Monsoon strength and variability have been and will continue to be crucial to the past and future prosperity of the region. With the emerging economies of China, Vietnam and India now adding to those of Japan, South Korea and Taiwan the importance of the region to the global economy has never been greater. Continuation of this growth is dependent on the climate and environment. Recent detailed climate reconstructions now show that the development and collapse of civilizations in both South and East Asia have been controlled in large measure by monsoon intensity. Modern technology now allows society to respond more effectively to environmental stresses, yet in the face of the destructive powers of typhoons or long duration droughts there is still little man can do when environmental catastrophe strikes.

As a result, understanding what controls the Asian monsoon and how it has changed in the past is important not only to scientists but also to the general population. In this book we present a multi-disciplinary overview of the monsoon for advanced students and researchers, spanning recent advances in atmospheric sciences, climatology, oceanography and geology. Finally we consider how the evolving monsoon has both helped and hindered the development of human civilizations since the Last Glacial Maximum, 20 000 years ago. The monsoon represents a large-scale seasonal reversal of the normal atmospheric circulation pattern. In this model, low-pressure systems develop in the tropics owing to rising hot air masses that cool and descend in the subtropics, which are thus characteristically arid regions. In contrast, summer heating of the Asian continent, especially around the Tibetan Plateau, generates low-pressure cells and thus summer rains in South and East Asia. In the winter a reversed high-pressure system is established, with dry, cold winds blowing out of Asia.

The links between the Tibetan Plateau and monsoon intensity have formed the basis of a long-running debate because this proposed relationship would appear to be one of the strongest examples of how the solid Earth, which is being continuously deformed and remodeled by plate tectonic forces, may be influencing the global climate system. The intensity of the modern monsoon likely reflects the fact that Tibet is the largest mountain chain seen on Earth for more than 500 million years and has correspondingly made a particularly large impact on the planet's atmospheric systems. Progress has been made in establishing links between the relatively slow growth of the plateau and monsoon strength, yet until the developing altitude of Tibet is better established and a truly long-scale climate history for the monsoon has been reconstructed it will remain impossible to test the linkages definitely. In particular, climatologists need an appropriate, long-duration sedimentary record dating back to the collision of the Indian and Asian plates that generated Tibet in the first place. In practice this means around 50 million years. Such a record exists in the oceans and continental margins around Asia, but has yet to be sampled.

While recognizing that the monsoon has strengthened over periods of millions or tens of millions of years, research focus over the past 10–15 years has demonstrated that not only does monsoon intensity vary dramatically on much shorter timescales, but that these are often linked to other parts of the global climate system. In particular, the detailed climate records now available for the past few million years show coherent, if sometimes lagged, development of the monsoon with the glaciation of the northern hemisphere. Clearly the monsoon cannot be studied in isolation from other systems, especially the oceanic–atmospheric systems of the North Atlantic (Gulf Stream and North Atlantic Deep Water) and the El Niño Southern Oscillation system of the Pacific Ocean. Indeed, it has been suggested not only that these systems control monsoon strength, but also that the monsoon can affect their evolution. A general pattern has emerged of summer monsoons being strong and winter monsoons generally weaker during warm, interglacial periods, and the reverse situation dominating during glacial times. As a result monsoon strength varies on the 21, 40 and 100 thousand year timescales that control periods of glacial advance and retreat. In detail, however, the situation is complicated by lags in the climate system that offset the response of the monsoon to solar forcing. In addition, there continues to be debate regarding how the monsoon differs in South and East Asia over various timescales. Current data suggest a generally coherent development between the two systems over millions of years but differences at the orbital and sub-millennial scale. Determining how and why they differ requires more high-resolution climate reconstructions from across the entire geographic range of the monsoon, involving both the "core area" of monsoon activity, such as the Bay of Bengal, and the "far-field" regions, such as the Sea of Japan and the Gulf of Oman, which may be more sensitive to modest changes in strength. Observations alone are not enough and a deep understanding of how the monsoon evolves and what the key controls are will require better climate models, ground-truthed with both oceanic and continental climate records.

The interactions of monsoon and society are a particularly fertile area of recent and future research. This field has developed as better climate records have been reconstructed over the past 8000 years or so. In particular the resolution permitted by ice cores and some high accumulation rate sediments in the oceans and lakes allows changes in monsoon intensity to be compared with human history. Indeed the ^{14}C dating used to constrain these records is the same method used to date archaeological sites, allowing a robust comparison to be made. Global warming, as a result of human activities, as well as natural processes, would tend to favor a stronger summer monsoon in the long term, yet in detail there is much potential complexity. Melting of the Greenland ice sheet may disrupt the overturn of waters in the North Atlantic and result in a cooling of that region. Comparison with similar natural events in the past suggests that such an event would result in weaker summer monsoons. Not only the strength of the monsoon can be affected by climate change but also its variability. Historical records indicate that the number and intensity of summer typhoons striking the densely populated coast of southern China have increased significantly over the past 200 years. If that trend were to continue, its economic and humanitarian effects could be disastrous.

Whatever part of the Earth we live in, the Asian monsoon is of significance to our lives and understanding of how the planet and society operates. Much work remains to be done in quantifying the monsoon and how it functions at a variety of timescales. Despite this great progress has been made in understanding this system. In this book we have attempted to synthesize what is now known and highlight those areas where significant research remains to be done.

PETER CLIFT
Aberdeen, UK
ALAN PLUMB
Cambridge, Massachusetts, USA

Acknowledgements

Clift would like to thank the following friends and colleagues for their generous help in putting this book together: Mark Altabet, An Zhisheng, David Anderson, Jon Bull, Doug Burbank, Stephen Burns, Kevin Cannariato, Marin Clark, Steve Clemens, Kristy Dahl, Dominik Fleitmann, Christian France-Lanord, Carmala Garzione, Liviu Giosan, Ananda Gunatilaka, Anil K. Gupta, Naomi Harada, Nigel Harris, Ulrike Herzschuh, David Heslop, Ann Holbourn, Yetang Hong, Tomohisa Irino, Hermann Kudrass, Wolfgang Kuhnt, Michinobu Kuwae, Peter Molnar, Delia Oppo, Dave Rea, Stephan Steinke, Ryuji Tada, Federica Tamburini, Ellen Thomas, Ruiliang Wang and Pinxian Wang. The idea for this book came from Clift's involvement in IGCP 476 "*Monsoons and Tectonics*," an international program supported by UNESCO and organized by Ryuji Tada.

Clift would also like to thank the Alexander von Humboldt Foundation for supporting his time in Bremen during a visiting fellowship at the Research Center for Ocean Margins (RCOM) and the Fachbereich Geowissenschaften at the Universität Bremen, when much of the writing was completed. Related research was completed thanks to financial support from the National Science Foundation (USA), the Natural Environment Research Council (UK), the Royal Society and the Carnegie Foundation for the Scottish Universities. Clift wishes to thank his wife Chryseis Fox for putting up with all the lost family time as a result of writing this book and the associated travel and research activities. Without her understanding and support this book would not have been possible.

We also wish to thank Bill Haxby and Suzanne Carbotte at the Lamont-Doherty Earth Observatory for their help with GeoMapApp, which is supported by the National Science Foundation.

We thank all the staff at Cambridge University Press who have helped with the production of this book, especially Matt Lloyd, Susan Francis, Dawn Preston, Denise Cheuk and Annette Cooper.

1

The meteorology of monsoons

1.1   Introduction

Monsoon circulations are major features of the tropical atmosphere, which, primarily through the rainfall associated with them, are of profound importance to a large fraction of the world's population. While there is no universally accepted definition of what constitutes a monsoon, there are some criteria that are widely accepted (see, e.g., the discussions in Ramage (1971), Webster (1987), and Neelin (2007)). Fundamentally, monsoonal climates are found where a tropical continent lies poleward of an equatorial ocean and are characterized by a strong seasonal cycle, with dry winters and very wet summers, and a reversal of wind direction from, in the dry season, the equatorward–easterly flow that is typical of most of the tropics to poleward–westerly flow after monsoon onset. Low-level flow from the ocean imports moisture onto the land to supply the rainfall there (although much of the rainfall within the monsoon system as a whole may actually fall over the neighboring ocean). In fact, in most monsoon systems this inflow includes strong cross-equatorial flow at low levels, from the winter to the summer hemisphere; however, this is not satisfied in all cases (such as the North American monsoon; Neelin (2007)). Indeed, given the differences in detail between different monsoon systems, even though they satisfy the most obvious criteria, it is inevitable that any attempt at definition will be imprecise, and even that classification of some regional meteorological regimes as monsoons may not be universally accepted.

The Asian–Indian Ocean–Australian monsoon system is, by some way, the most dramatic on the planet in terms of its intensity and spatial extent, but there are other regions of the globe, specifically North and Central America, and West Africa, that display similar characteristics and are thus classified as monsoons. It is important to recognize at the outset that, despite these regional classifications, the monsoons form part of the planetary-scale circulation of the tropical atmosphere: they are influenced by, and in turn influence, the global circulation. Accordingly, we shall begin this overview with a brief review of the "big picture" of the tropical circulation, which will lead into a more focussed discussion of the Asian–Indian Ocean–Australian monsoon system.

1.2   Meteorology of the tropics

1.2.1   Observed zonal mean picture

A good starting point for understanding the general circulation of the global atmosphere is to look at the zonally (i.e., longitudinally) averaged circulation in the meridional (latitude-height) plane. Since the circulation varies seasonally (an essential fact of monsoon circulations) it is better to look at seasonal, rather than annual, averages. In turn, the atmosphere exhibits interannual variability – it is a matter of basic experience that one year's weather differs from the last, and this is especially true in the tropics – and so, in a general overview such as this, we shall look not at individual years, but at *climatological* averages, i.e., averages over many summers or winters, which show the normal picture for that season.

Figure 1.1 shows the climatological distribution of mean zonal wind and temperature for the two solstice seasons DJF (December through February) and JJA (June through August). The dominant features of the zonal wind distribution are two westerly subtropical jets straddling the equator at altitudes of about 12 km (near 200 hPa pressure). The core of the stronger jet is located at about 30° latitude in the winter hemisphere, while that of the weaker jet is at 40–50° latitude in the summer hemisphere. Within the deep tropics, the zonal wind is easterly, though mostly weak, all the way down to the surface. Outside the tropics, at latitudes greater than about 30°, the mean surface winds are westerly.

Several features of the mean temperature distribution are worthy of note. Temperature generally decreases rapidly through the troposphere up to the tropopause whose mean altitude varies from about 17 km in the tropics down to around 8 km (near 400 hPa) at the poles. Above, temperature increases, or decreases more slowly, with altitude through the stratosphere. As will be seen in Figure 1.3, almost all atmospheric water (along with most dynamical processes relevant to surface weather and climate) is located in the troposphere. Within the troposphere, temperature decreases systematically poleward from a broad maximum centered in the summer tropics. Note, however, the weak temperature gradients between the two subtropical jets, which contrast with the strong gradients in middle latitudes, poleward of the jet cores.

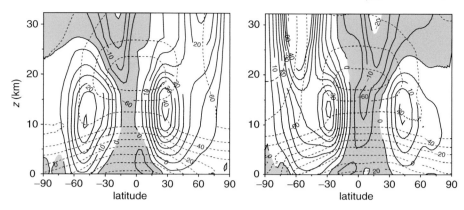

Figure 1.1 Climatological zonal mean zonal wind (solid; ms^{-1}) and temperature (dashed; °C) for (left) December–February and (right) June–August. Contour intervals are 5 ms^{-1} and 10°, respectively; easterly winds are shaded. The data are averaged on pressure surfaces; the height scale shown is representative. Data provided by the NOAA-CIRES Climate Diagnostics Center, Boulder, Colorado, through their website at www.cdc.noaa.gov/.

Continuity of mass requires that the zonal mean circulation in the meridional plane be closed, so that northward and vertical motions are directly linked. A convenient way to display the meridional circulation on a single plot is to show the mass streamfunction χ, which is done in Figure 1.2 for the two solstice periods. The mean northward and upward velocities (v,w) are related to the mass streamfunction χ through

$$v = \frac{-1}{2\pi\rho a\cos\varphi}\frac{\partial\chi}{\partial z}; \qquad w = \frac{1}{2\pi\rho a^2\cos\varphi}\frac{\partial\chi}{\partial\varphi},$$

where ρ is the density, a is the Earth's radius and ϕ the latitude. The velocities are thus directed along the χ contours, with mass flux inversely proportional to the contour spacing. In this plane, the mean circulation is almost entirely confined to the tropics. This tropical cell is known as the Hadley circulation, with upwelling over and slightly on the summer side of the equator, summer-to-winter flow in the upper troposphere, downwelling in the winter subtropics, and winter-to-summer flow in the lower troposphere. The latitude of the poleward edge of the cell coincides with that of the winter subtropical jet. There is a much weaker, mirror-image, cell on the summer side of the equator. Around the equinoxes, the structure is more symmetric, with upwelling near the equator and downwelling in the subtropics of both hemispheres.

The distribution of atmospheric moisture is shown in Figure 1.3. Humidity is expressed in two forms: *specific humidity*, the amount of water vapor per unit

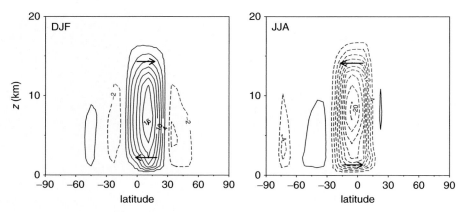

Figure 1.2 Climatological zonal mean overturning streamfunction ($10 \, \text{kg s}^{-1}$ for December–February (left) and June–August (right). Solid contours denote positive values, dashed contours are negative; the zero contour is not plotted. The meridional flow is directed along the streamfunction contours, clockwise around positive cells, anticlockwise around negative cells, as indicated for the dominant cells by the arrows on the plots. The magnitude of the net mass circulation around each cell is equal to the value of the streamfunction extremum in the cell. Data provided by the NOAA-CIRES Climate Diagnostics Center, Boulder, Colorado, through their website at www.cdc.noaa.gov/.

mass of air, conventionally expressed as g kg^{-1}, and *relative humidity*, the ratio of specific humidity to its saturation value (the value in equilibrium with liquid water at the ambient temperature and pressure). On this zonally and climatologically averaged view, the near-surface relative humidity varies remarkably little across the globe, being mostly between 65 and 85%. The driest surface regions are near the poles, and in the desert belt of the subtropics. There is a general decrease of relative humidity with height, a consequence of the drying effects of precipitation in updrafts followed by adiabatic descent; the regions of subsidence on the poleward flanks of the Hadley circulation are particularly undersaturated. The zonally averaged specific humidity is as large as $17 \, \text{g kg}^{-1}$ near the surface just on the summer side of the equator, decaying to less than $1 \, \text{g kg}^{-1}$ in high latitudes and in the upper troposphere and above. Indeed, the variation of specific humidity is much greater than that of relative humidity, indicating that the former primarily reflects variations of saturation vapor pressure, which has a very strong dependence on temperature (expressed as the Clausius–Clapeyron relationship; see, e.g., Bohren and Albrecht (1998)). Thus, the highest specific humidities are found where the atmosphere is warmest: at low altitudes in the tropics.

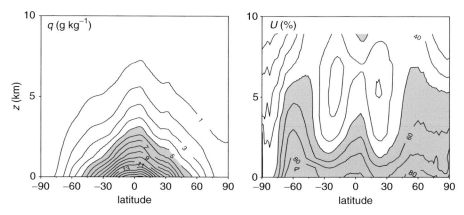

Figure 1.3 Climatological annual- and zonal-mean specific humidity (left, g kg^{-1}; values greater than 5 g kg^{-1} are shaded) and relative humidity (right, %; values greater than 50% are shaded). Data provided by the NOAA-CIRES Climate Diagnostics Center, Boulder, Colorado, through their website at www.cdc.noaa.gov/.

1.2.2   Dynamical and thermodynamical constraints on the circulation

At first sight, some of the characteristics of the zonally averaged atmosphere may seem puzzling. Ultimately, what drives the atmospheric circulation is the spatial variation of the input of solar energy (per unit surface area) into the atmosphere, which generally decreases monotonically from a maximum in the summer tropics to minima at the poles, yet the meridional circulation is not global in extent. Rather, it terminates at the edge of the tropics where the subtropical jets are located, and there is a distinct contrast between, on the one hand, the tropical region between the jets, characterized by weak horizontal temperature gradients, the strong Hadley circulation, and easterly winds and, on the other hand, the extratropical regions of strong temperature gradients, weak mean meridional flow, and westerly winds poleward of the jets. There is no such sharp distinction in the external forcing.

The most important controlling factor separating the meteorology of the tropics from that of middle and high latitudes is the Earth's rotation. Consider air rising near the equator, and turning toward the winter pole as seen in Figure 1.2. If for the moment we consider zonally symmetric motions, the air aloft (where frictional losses are utterly negligible) will conserve its absolute angular momentum – angular momentum relative to an inertial reference frame, which includes components associated with the planetary rotation as well as with relative motion – as it moves. As air moves away from the equator and thus closer to the rotation axis, the planetary component decreases; consequently, the relative motion must increase. The further poleward the air moves, the more dramatic the effects of rotation become, just because of the geometry of the sphere. Thus, the winds would become increasingly westerly (eastward) with latitude, and dramatically so: $58\,\mathrm{ms}^{-1}$ at $20°$, $134\,\mathrm{ms}^{-1}$ at $30°$, $328\,\mathrm{ms}^{-1}$ at $45°$. In fact, the westerly wind would have to become infinite at the pole. At some point, the atmosphere cannot sustain equilibrium with such winds. Consequently the poleward circulation must terminate at some latitude; exactly where is determined by many factors, most importantly a balance between the strength of the external forcing and the effective local planetary rotation rate (Held and Hou, 1980; Lindzen and Hou, 1988). These termination latitudes mark the poleward boundaries of the Hadley circulation, and the latitude of the subtropical jet. (In reality, the jets are weaker than this argument would imply; processes we have not considered here – most importantly, angular momentum transport by eddies – allow the air to lose angular momentum as it moves poleward.)

Rotational effects are manifested in the balance of forces through the Coriolis acceleration which, for the large-scale atmospheric flow, is more important than the centripetal acceleration. In general, the vector Coriolis acceleration is $2\boldsymbol{\Omega} \times \mathbf{u}$, where $\boldsymbol{\Omega}$ is the vector planetary rotation rate and $\mathbf{u}$ the vector velocity. However, the atmosphere is so thin that the vertical component of velocity is necessarily much smaller than the horizontal components and, in consequence, the important components of acceleration can be written as $f\hat{\mathbf{z}} \times \mathbf{u}$, where $f = 2\Omega\sin\varphi$, the Coriolis parameter, is just twice the projection of the rotation rate onto the local upward direction $\hat{\mathbf{z}}$. At low latitudes f, and hence the influence of planetary rotation, is weak, thus permitting the Hadley circulation to exist there. This fact also implies that pressure must approximately be horizontally uniform, just as the surface of a pond must generally be flat (ponds typically being much too small for planetary rotation to matter). Since, in hydrostatic balance, the pressure at any location is just equal to the weight of overlying air per unit horizontal area, and density depends on temperature, the horizontal temperature gradients there must also be weak, as is observed in the tropical atmosphere (Figure 1.1). In fact, the fundamental role of the Hadley circulation is to maintain this state. Thus, the existence of a separation of characteristics between the tropical and extratropical regions of the atmosphere is, in large part, a consequence of planetary rotation.

These, and essentially all other, atmospheric motions derive their energy ultimately from the input of solar energy or, more precisely, from the differential input between low and high latitudes, which creates internal and potential energy within the atmosphere, a portion of which is then converted into the kinetic energy of atmospheric winds. For a compressible atmosphere in hydrostatic balance, internal and potential energy are closely related to each other;

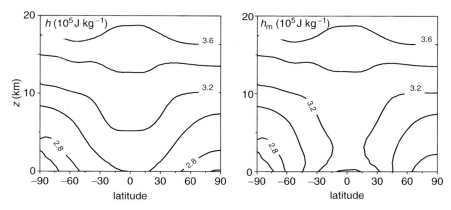

Figure 1.4 Climatological annual- and zonal-mean meridional distribution of (left) dry static energy and (right) moist static energy ($J\,kg^{-1}$). Data provided by the NOAA-CIRES Climate Diagnostics Center, Boulder, Colorado, through their website at www.cdc.noaa.gov/.

accordingly, it is conventional to combine them into a quantity known as *dry static energy*, which, per unit mass of air, is

$$h = c_p T + gz,$$

where z and T are altitude and temperature, g is the acceleration due to gravity, and c_p is the specific heat of air at constant pressure. The annual- and zonal-mean distribution of h is shown in the left frame of Figure 1.4.

Just as planetary rotation constrains horizontal motion, so thermodynamic effects and gravity restrict vertical motion. Dry static energy increases with height at all latitudes. Therefore, for near-equatorial air in the upwelling branch of the Hadley circulation to move from the surface up to the upper troposphere (Figure 1.2), its dry static energy must increase. Moreover, in practice what appears in Figure 1.2 as a broadscale, slow, upwelling is in fact the spatial and temporal average of much more rapid motion within narrow convective towers; in such towers, air typically moves from surface to tropopause in an hour or so. Radiation cannot provide the implied diabatic heating: it is much too weak and, besides, radiation is generally a cooling agent in the tropics. However, as was evident in Figure 1.3, tropical surface air is very moist, and the near-equatorial upwelling is thus characterized by saturation, condensation and intense rainfall. Condensation is a major contributor to the thermodynamic balances. Many treatments focus on the thermodynamics of dry air, but add adiabatic heating equal to $-L \times dq/dt$ per unit mass per unit time, where L is the enthalpy (latent heat) of vaporization and q the specific humidity, so that $-dq/dt$ is the rate of condensation per unit mass of air.

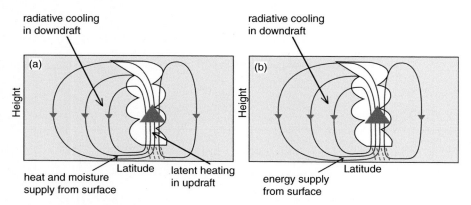

Figure 1.5 Schematic depiction of the energetics of the Hadley circulation. In frame (a), moisture is treated as a source of static energy; in frame (b), it is treated as an integral component of the atmospheric static energy. (See text for discussion.)

The thermodynamic driving of the tropical meridional circulation can thus be described as illustrated in Figure 1.5(a). Air is supplied with heat and moisture in the near-surface boundary, as it flows across the warm tropical ocean (and the stronger the low-level wind, the more turbulent the boundary layer and the greater the evaporation from the ocean surface). Ascent, requiring increasing h, is facilitated by the release of latent heat associated with condensation of water vapor, and consequent precipitation, in the relatively concentrated updraft. The compensating loss of energy occurs in the downwelling region. Descending air tends to warm adiabatically; this warming must result in enhanced emission of thermal radiation. So the picture of Figure 1.5(a) is one of heating in the updrafts, cooling in the downdrafts.

The foregoing description, although commonly found in meteorology texts until quite recently, is imperfect for many reasons. For our purposes, the one important reason is that it misleads us into believing that the underlying driver for such circulations is the latent heat release consequent on precipitation in the updraft, whereas the energy of the circulation is, in fact, supplied ultimately from the warm underlying ocean. More completely, it is the contrast between energy gain in the warm surface boundary layer and the radiative energy loss at lower temperature (in the cooler middle and upper troposphere) in the downwelling. Thus, this kind of circulation is just a classical Carnot heat engine[1] (Emanuel, 1986). To appreciate this fact, we need to recognize that latent heat

---

[1] This statement is true of many tropical circulation systems. With a modest change of geometry, Figure 1.5 could equally well depict a hurricane or, as we shall see, a monsoon circulation.

release does not constitute an external source of energy. Rather, the process of condensation is internal to an air parcel, and a better way of treating its effects is to include moisture directly in the definition of *moist static energy*

$$h_m = c_p T + gz + Lq$$

per unit mass (e.g., Emanuel, 2000; Holton, 2004). The climatological annual- and zonal-mean distribution of h_m is shown in the left frame of Figure 1.4. The greatest difference between h_m and h is, not surprisingly, in the tropical lower atmosphere where q is greatest. For our purposes, the most important feature is the elimination of the vertical gradient in the deep tropics: when the contribution of moisture to entropy is properly taken into account, therefore, there is no need to invoke heating in the updraft, since the moist entropy of the air does not change in the updrafts. From this, thermodynamically more consistent, viewpoint there is thus no heating (i.e., no external tendency to increase energy) in the updraft: moisture is lost (to precipitation) but the consequent temperature change is such as to preserve h_m. Instead, as depicted in Figure 1.5(b), the energy source driving the circulation is located at the surface, where the crucial role of the supply of both sensible and latent heat from the warm ocean now becomes very explicit. We are thus led to recognize the moist static energy of boundary layer air as a key factor in understanding tropical circulations.

1.2.3   Longitudinal variations in tropical meteorology

The distributions of surface pressure over the tropics in the solstice seasons are shown in Figure 1.6. Pressure variations in the tropics are relatively weak, typically a few hPa, as compared with typical variations of 10–20 hPa in extratropical latitudes. A continuous belt of low pressure spans all longitudes, mostly located near the equator over the oceans but displaced into the summer hemisphere over the continents. An almost continuous belt of high surface pressure characterizes the subtropical region around 30° latitude, but in the summer hemisphere the high pressure band is interrupted by continental lows, leaving high pressure centers over the oceans.

The low-level winds (shown for the 850 hPa surface, near 1 km altitude, in Figure 1.7) are dominated by the north-easterly and south-easterly Trade winds in the northern and southern hemisphere, respectively. This general pattern is, however, modulated by features that reflect the pressure distribution. The low-level winds converge into the band of low pressure; in regions, especially oceanic regions, where this band forms longitudinally elongated features, it is known as the Intertropical Convergence Zone, or ITCZ. The Pacific and Atlantic Ocean ITCZs are located north of the equator throughout the year but in this respect, as in many others, the circulation over the Indian Ocean behaves differently.

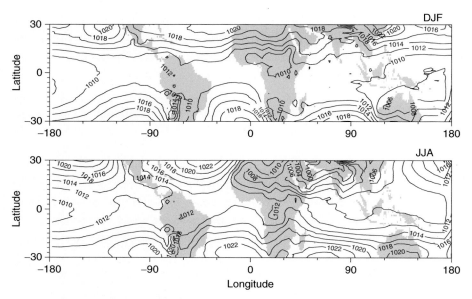

Figure 1.6 Climatological (long-term average) surface pressure (hPa) over the tropics in (top) December–February and (bottom) June–August. Data provided by the NOAA-CIRES Climate Diagnostics Center, Boulder, Colorado, through their website at www.cdc.noaa.gov/.

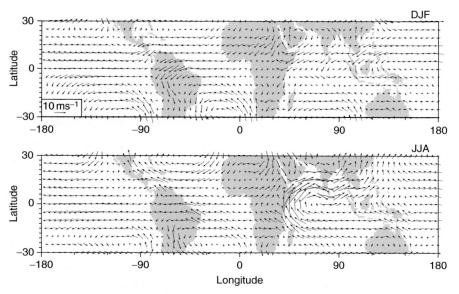

Figure 1.7 Climatological mean winds at 850 hPa (near 1 km altitude) in (top) December–February and (bottom) June–Aug. The scale for the arrows is shown at the lower left of the top plot. Data provided by the NOAA-CIRES Climate Diagnostics Center, Boulder, Colorado, through their website at www.cdc.noaa.gov/.

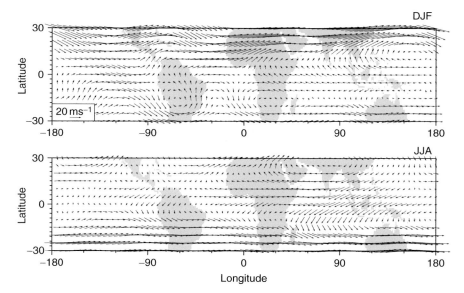

Figure 1.8 As Figure 1.7, but for the 200 hPa pressure level.

The Indian Ocean ITCZ is south of the equator in southern hemisphere summer, but in northern summer there is no clear oceanic convergence zone; rather than converging near the equator, the low-level flow crosses the equator over the western part of the ocean, becoming south-westerly north of the equator. This dramatic feature is the inflow to the South Asian monsoon system, which will be discussed in more detail in Section 1.3. The large surface anticyclones (of high pressure) noted on Figure 1.6 are regions of flow *divergence*; the low-level flow spirals out of these mostly oceanic features to form the Trade winds and ultimately to feed the convergence zones.

The climatological upper-level tropical winds (on the 200 hPa pressure surface, near 11 km altitude) are shown in Figure 1.8. The winds are stronger aloft; they are mostly easterly very close to the equator, and strongly westerly in the subtropics; thus, the zonally averaged picture presented in this section is fairly representative of most longitudes. A few large, strongly divergent anticyclones[2] are evident in the region between the easterlies and westerlies, in fact almost directly over the regions of low-level convergence noted previously, most markedly over the summer continents. There is also strong upper-level outflow from the oceanic ITCZs, but this is somewhat masked on Figure 1.8 by the strong nondivergent part of the flow there. It is here that the large-scale

---

[2] Anticyclones are evident in the wind plots as clockwise or anticlockwise circulations north or south of the equator, respectively.

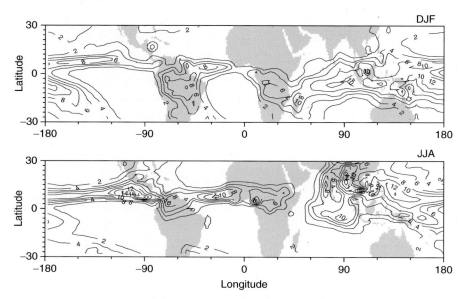

Figure 1.9 Climatological mean rainfall rate (mm day^{-1}) over the tropics in (top) December–February and (bottom) June–August. Data provided by the NOAA-CIRES Climate Diagnostics Center, Boulder, Colorado, through their website at www.cdc.noaa.gov/.

tropical ascent is taking place, fed by low-level convergence into surface low pressure and spiralling out in the upper tropospheric anticyclones.

These tropical regions of strong, deep ascent are, of course, characterized by heavy rainfall (Figure 1.9); since the saturation-specific humidity at the outflow level is no more than a small percentage of the specific humidity of the surface air, nearly all the moisture converging at low levels condenses in the updraft to fall as precipitation. In consequence, annual mean rainfall in these regions is as much as several meters.

1.2.4   Location of the convergence zones

Why are the convergence regions, with their associated deep convection and intense rainfall, located where they are? Over the oceans, the answer is straightforward: to a large degree, they follow the warmest water. Figure 1.10 shows the climatological distribution of surface temperature in the solstice seasons. The tropical oceans are generally warm, of course; on the large scale, the warmest waters are located consistently in the western tropical Pacific Ocean, and across what is sometimes referred to as the maritime continent, the region lying between the Pacific and Indian Oceans, and occupied by Indonesia, Melanesia, and other island groups. In this region, where sea-surface temperature (SST) can reach 30 °C, the locus of the warmest waters migrates

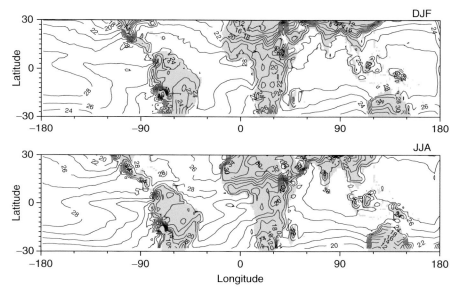

Figure 1.10 Climatological mean surface temperature (°C) in (top)
December–February and (bottom) June–August. Data provided by the NOAA-CIRES
Climate Diagnostics Center, Boulder, Colorado through their website
at www.cdc.noaa.gov/.

a little with the seasons, toward the summer hemisphere. Elsewhere, the
warmest waters occur in zonally extensive, but rather narrow, bands which
are located mostly north of the equator throughout the year in the Pacific and
Atlantic Oceans, but south of the equator in southern summer in the Indian
Ocean. Strikingly, in a narrow band a few degrees of latitude wide along the
equator, and especially in the eastern Pacific and Atlantic Oceans, the sea surface
is much colder, a consequence of the upwelling there of cold water from depth.

Comparing the distributions of low-level winds and rainfall shown in
Figures 1.7 and 1.9 with that of surface temperature in Figure 1.10, it is clear
that the oceanic high-rainfall convergence zones are collocated with the warmest
water. The atmospheric circulation in these regions is thus controlled by the
oceans. The ocean surface in these regions is not only warm it is also (to state
the obvious) wet and moisture supply to the atmospheric boundary layer is
uninhibited. Because of the near-exponential dependence of the saturation
specific humidity on temperature, the boundary layer becomes extremely moist,
as well as warm, and thus acquires large moist static energy h_m. High boundary
layer h_m is the primary condition for atmospheric convection; accordingly, the
correspondence between high SST and deep atmospheric convection, intense
rain, and low-level flow convergence is straightforward.

One consequence of the strong east-west variations in equatorial SST, especially in the Pacific Ocean, is that the atmospheric circulation comprises not only the Hadley circulation, with its overturning in the latitude–height plane, but also an east–west overturning in much of the tropics. The latter circulation is most evident in northern winter, when it is manifested in Figures 1.7 and 1.8 as low-level equatorial easterly winds overlain by westerlies. The strongest such circulation is in the Pacific basin, where it is known as the Walker circulation, in which air rising over the warm maritime continent spreads westward along the equator, subsides and then returns as a low-level easterly flow. There is, in fact, a strong interplay between the atmospheric Walker circulation and the SST distribution in the equatorial Pacific Ocean: the Walker circulation is controlled by the longitudinal gradient of SST while, in turn, the SST distribution is strongly influenced by the wind stress consequent on the low-level atmospheric flow. Fluctuations in this coupled system manifest themselves as the El Niño–La Niña cycle in the ocean and as the Southern Oscillation in the atmosphere (and, in fact, the coupled phenomenon is now widely known by the concatenated acronym ENSO).

Over the continents, the seasonal shift of the convergence zones is much more marked than over the oceans. There appear to be two types of continental behavior: the region of high rainfall in the interior of Africa migrates relatively smoothly back and forth with the seasons between extremes of about 15° S in southern summer (reaching 20° S in Madagascar) and around 10° N in northern summer. At the equinoxes, the rain band passes the equator; thus, while rainfall peaks once per year (in the wet season of local summer) north and south of the equator, there are two wet seasons per year (at the equinoxes) in the equatorial belt. Similar behavior is also seen in tropical America: it thus appears to be characteristic of continents that span the equator.

In other areas, where tropical land lies poleward of ocean, the seasonal transition from dry to wet to dry is more dramatic. The clearest example of this is the South Asian–Indian Ocean–Australian monsoon system, on which we shall focus in what follows.

1.3   The Indian Ocean monsoon system

The distribution of rainfall over the Indian Ocean region for the four seasons is shown in Figure 1.11; low-level winds over the same region are plotted in Figure 1.12. Through the course of a year, the bulk of the rainfall over the Indian Ocean region migrates systematically from a broad region extending across the maritime continent and the southern equatorial Indian Ocean in northern spring, moving onto the southern and south-eastern Asian continent

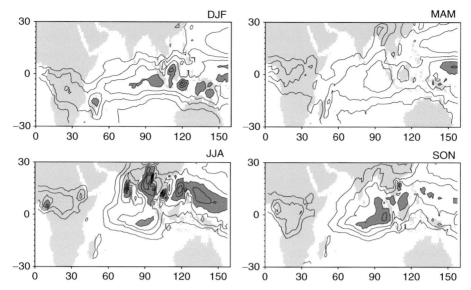

Figure 1.11 Climatological mean rainfall rate for the four seasons. The contour interval is 2.5 mm day^{-1}; heavy shading denotes rainfall greater than 5–10 mm day^{-1}.

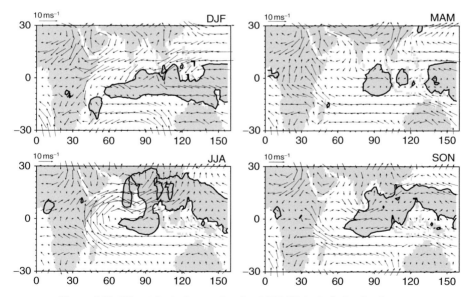

Figure 1.12 Climatological mean low level (850 hPa) winds for the four seasons. The scale for the wind arrows is shown at the top left of each plot. The heavy lines mark regions of seasonal mean rainfall in excess of 7.5 mm day^{-1}.

and nearby oceanic regions in summer, returning southward in northern autumn, reaching as far as northern Australia in the east and Madagascar in the west, in southern summer. During the most active period of northern summer, the intense rainfall maximum near the Asian coast is accompanied by vigorous deep convection and the concomitant net upwelling. Warm, moist air is supplied from the south-west by a strong inflow originating in the easterlies of the southern tropics, crossing the equator in the western ocean and curving north-eastward across the Arabian Sea, and from the south-east across the maritime continent. It is thus clear that this system is not at all local, but a planetary-scale phenomenon. (Indeed, during northern summer, the implied meridional overturning circulation is the dominant contribution to the global Hadley circulation shown in Figure 1.2.) The strong cross-equatorial flow, a feature of most (if not all, see Neelin (2007)) monsoon circulations, is especially strong over the western ocean where in fact it forms a northward jet on the eastern flank of the East African highlands (Findlater, 1969). The cross-equatorial flow over the western ocean is thus concentrated below the peaks of and alongside the mountains. In fact, much of the northward flow crossing the equator in the eastern ocean, in the vicinity of the maritime continent, is also concentrated adjacent to and between the mountains there.

Climatological conditions prior to and following the onset of this large-scale flow are illustrated in Figure 1.13. In May, while some northward cross-equatorial flow is becoming established at the East African coast, the flow is weak and it does not yet extend toward the Indian subcontinent. Flow onto the continent has, however, become established over south-east Asia by this time. The surface waters of the northern Indian Ocean have warmed, especially in the Arabian Sea and the Bay of Bengal. The rain belt (not shown) is beginning to shift north of the equator east of 90° E, but not further west. Hot, dry

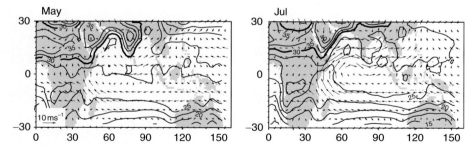

Figure 1.13 Climatological 1000 hPa air temperature and 850 hPa winds over the Indian Ocean region in (left) May and (right) July. Contour interval 2.5 °C; heavy contour is 30 °C. The arrow scale for both frames is shown in the lower left of the left frame.

conditions extend from the Sahara across Arabia and India. In the interior of the
Indian subcontinent, temperatures exceed 40 °C, but deep convection does not
occur, in part because of the dryness of the low level air, such that the
near-surface moist static energy is actually less than that over the somewhat
cooler oceans to the south, and because of broad-scale subsidence aloft. However,
this state of affairs eventually breaks down, accompanied by the almost simultaneous onset of deep convection and rainfall across the northern Indian Ocean
and the continent south of about 20° N, and of the large-scale summertime
circulation. The suddenness with which the circulation is established is evident
in Figure 1.14, which shows the winds picking up over the Arabian Sea over
an interval of 2–3 weeks. With onset, temperatures in the subcontinent drop
markedly (Figure 1.13) as cooler but more humid air is advected inland. The
precipitation maximum over the Indian Ocean migrates quite rapidly from
near and south of the equator to be centered near 10° N, although equatorial
precipitation does not cease with monsoon onset. The maximum moves further
north over the following month before receding, more gradually than during
onset, in October to November. Surface temperatures off the coast also drop,
partly in response to enhanced evaporation in the strong winds.

Despite the rather systematic onset of the large-scale flow evident in
Figure 1.14, the appearance of monsoon rains is not simultaneous over southern

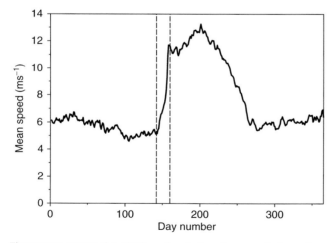

Figure 1.14 Seasonal evolution of an index of mean wind speed over the Arabian Sea.
The index is calculated as the square root of specific kinetic energy, averaged over the
area 5° S to 20° N, 50° E to 70° E. The day number is a nominal day of year (day 1 is
nominally Jan 1), shifted with respect to onset date (the data are composited with
respect to monsoon onset). The vertical dashed lines indicate days number 142 and
160. Data courtesy of William Boos.

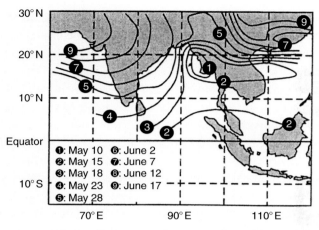

Figure 1.15 Climatological date of the onset of monsoon precipitation, as indicated by the mean location of the 220 Wm^{-2} contour of outgoing long-wave radiation (OLR) observed from space. (Values of OLR as low as this are indicative of cold, high cloud and thus of deep convection and intense rainfall.) (Webster *et al.*, 1998).

and south-eastern Asia. While the actual progression varies somewhat from year to year, on average the rains typically first appear in the eastern Bay of Bengal and Indochina in mid-May, and extend westward and northward from there across the Indian subcontinent and south-eastern Asia over the course of the following month, as illustrated in Figure 1.15. Thus, the actual date of onset, either climatologically or in a given year, is sensitive to the definition used (e.g., Fasullo and Webster, 2003).

Until recently, detailed mapping of monsoon precipitation has been problematic, because of the relative sparseness of high quality rain-gage observations and the highly heterogeneous nature of rainfall. Now, however, high-resolution space-based observations have greatly improved the situation, and maps of mean rainfall rates for June to August from three sources are shown in Figure 1.16. While there are significant differences in intensity between the different data sources (and note that the years making up the averages differ from frame to frame), the benefit of and necessity for high resolution is obvious. The greater part of the precipitation in fact occurs over the ocean, especially the northern and eastern Bay of Bengal and the eastern South China Sea. Rainfall maxima over land are clearly linked to topographic features (Xie *et al.*, 2006): along the Western Ghats, the foothills of the Himalaya, the mountain ranges on the west coast of Burma, Indochina and the west coast of Luzon in the Philippines. These maxima are embedded within the weaker, but still very substantial, region of rainfall extending across the Indian subcontinent and south-eastern Asia evident in both Figures 1.11 and 1.16.

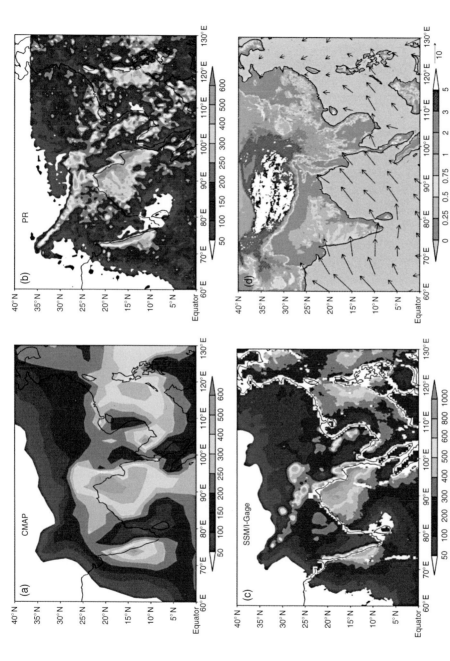

Figure 1.16 Summertime rainfall over the northern Indian Ocean and South Asia. Frames (a)–(c) show June–August mean rainfall rates (mm per month) from different data sources: (a) the Climate Prediction Center Merged Analysis of Precipitation (CMAP; 2.5° resolution, coverage 1979–2003); (b) high resolution data from the space-based Tropical Rainfall Measurement Mission Precipitation Radar (1998–2004) and (c) merged high resolution data from the space-based Special Sensor Microwave Imager and land-based observations (1987–2003). Surface topography (km) is shown in frame (d). (Xie *et al.*, 2006). See color plate section.

19

1.3.1   Intraseasonal variability of the monsoon

Within the wet monsoon season, rainfall intensity fluctuates considerably from day to day and from week to week. There are well defined active and break cycles, each of which may last from a few days to a few weeks (e.g., Webster *et al.*, 1998). During the active cycle, the large-scale monsoonal flow pattern intensifies, rainfall rates increase across much of southern Asia and the northern Indian Ocean and decrease south of the equator. However, just as during onset, a break period does not necessarily imply weak rainfall everywhere. In fact, the rainbands tend to drift northward during the active-break cycle (e.g., Sikka and Gadgil, 1980) such that heavy rains may be falling over the Himalaya foothills while rains subside over much of the rest of the region.

There has been much discussion in the literature about the relationship between these cycles of the Asian monsoon and the Madden–Julian Oscillation (MJO). The MJO (Madden and Julian, 1972, 1994) is a persistent, large-scale and dominant feature of tropical meteorology, which propagates systematically eastward around the tropics with a variable period of around 30–50 days. It has its strongest manifestations in the Indian and Pacific Oceans, but is weakest in northern summer when it might interact with the Asian monsoon. While fluctuations of the Asian monsoon system do have significant variance in the MJO period range (e.g., Yasunari, 1980, 1981) the relationship between the two is not entirely clear (Webster *et al.*, 1998). There does, however, appear to be a clear connection between the MJO and the Australian monsoon (Hendon and Liebmann, 1990).

On shorter timescales, rainfall is modulated by the passage of monsoon depressions. In fact, much of the rain that falls during the monsoon season is associated with these systems. An example is illustrated in Figure 1.17. Typically, four to five of these depressions form in the Bay of Bengal during a monsoon season (Rao, 1976); they propagate generally westward or north-westward over the subcontinent over the course of a few days, bringing intense rain, mostly to the west and south-west of the storm center.

1.3.2   Interannual variability

Figure 1.18 illustrates the year-to-year variation of annual mean rainfall for India and for northern Australia. Annual mean rainfall in both regions is dominated by the monsoon rains. Perhaps the first remark to be made, to echo Webster *et al.* (2002), is that interannual variability is, especially for the all-India composite, remarkably weak, in contrast to the more dramatic year-to-year variations that characterize rainfall records across much of the tropics. Nevertheless, year-to-year variations are apparent, the most dominant of which is a two-year cycle in which a wet year is followed by a dry one, and for which the Asian and Australian monsoons appear to be correlated, i.e., a strong South

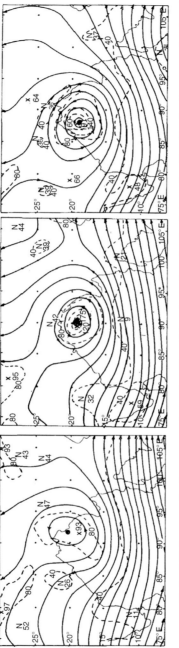

Figure 1.17 An example of a monsoon depression. The solid contours show low-level streamfunction (the dominant, rotational, component of the flow is along these contours, anticlockwise around the storm center). From left to right, the frames are at 24 h intervals beginning 1200 GMT 5 July 1979. (From Sanders (1984)).

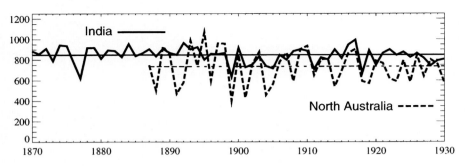

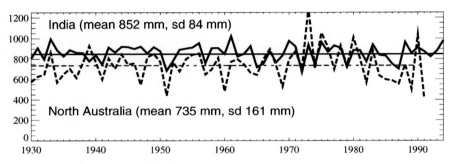

Figure 1.18 Annual-mean rainfall (mm) showing all-India rainfall (solid) and north Australia rainfall (dashed). (Webster *et al.*, 2005).

Asian monsoon tends to be followed, in the ensuing northern winter, by a strong Australian monsoon (Meehl, 1994). Indeed, the two monsoons appear to be intimately linked with Indian Ocean temperatures, with strong monsoon rainfall associated with warm sea-surface temperatures in the preceding northern winter (Harzallah and Sadourny, 1997; Clark *et al.*, 2000).

The dominant signal of interannual variability affecting much of the tropics, especially in the Pacific basin, is that associated with ENSO. El Niño years (identified by anomalously warm surface temperatures in the eastern equatorial Pacific Ocean) are associated with dry conditions in the west Pacific–Indian Ocean region, and this is strongly evident in South Asian monsoon rainfall (Shukla and Paolina, 1983; Torrence and Webster, 1999). Conversely, in the opposite phase of ENSO, La Niña years are usually associated with above-average rainfall.

1.4   Theory of monsoons

1.4.1   General considerations

Our understanding of the meteorology of monsoons, and of the factors that control them, has many facets, and is still far from complete. At the very simplest level, the monsoon system is very much like the Hadley circulation illustration of Figure 1.5: air is supplied with energy at the hot, moist surface, driving an overturning circulation and compensating energy loss in the warm downwelling region. However, real-world monsoons are much more complex than this. For one thing, the underlying surface is heterogeneous: monsoon circulations encompass both oceanic and continental regions. This fact appears to be crucial; the nature, and seasonal variability, of climates and atmospheric circulations in continental and oceanic regions are very different. More to the point, it has long been recognized that it is the land–sea contrast that is at the heart of driving the whole monsoon system. Further, the intense rainfall over land that is characteristic of monsoons, and which is intimately connected to the upward branch of the circulation, cannot be sustained without import of moisture from the ocean, and so there is an inherent two-way feedback between the circulation and the precipitation. Also, monsoon circulations are three-dimensional (i.e., they are longitudinally localized), which makes their dynamics different from that of a simple, zonally symmetric, Hadley circulation.

Just as in Figure 1.5, it is energy input into the atmosphere that drives the circulation, and most of this input is from the surface. Ultimately, of course, all the energy input is from the sun, but the way this energy finds its way into the atmosphere is very different for continental and oceanic regions. Land has a small heat capacity, so that whatever energy the surface receives from the sun (and via thermal radiation from the atmosphere and clouds) is transferred almost immediately to the overlying atmosphere. Not only does the ocean have a much greater heat capacity, but also currents and mixing within the ocean exert a strong influence on SST; accordingly, there is not such a direct, nor immediate, relationship between the input of energy into the ocean and its transfer into the atmosphere. In the subtropics, therefore, as the local downward flux of solar radiation increases with the onset of summer, the surface fluxes over land increase dramatically, thus (other things being equal) increasing h_m, the moist static energy of the boundary layer, while h_m over the ocean is controlled by the more slowly evolving sea-surface temperatures. In early summer, air over the continent is dry; accordingly, almost all of the input of solar energy goes into internal energy: temperature increases, to very high values (cf. Figure 1.10). In the early stages, continental h_m is still less than that over the adjacent ocean, despite its high temperature, because of its low specific humidity. Consequently, convection continues to be located over the ocean, and so there is little change in the large-scale atmospheric circulation. Eventually, however, h_m over the land begins to exceed that over the warmest SSTs; at this stage, convection is favored over the land (it may or may not collapse over the ocean); simultaneously, the associated net upward motion induces inflow, supplying moisture to the continental lower atmosphere, thus feeding moist convection and intense rain. (The input of moisture reduces continental temperatures, since now a significant fraction of moist static energy is in latent form.)

The observed monsoon circulation is, in fact, very much in accord with what theory predicts, given the observed rainfall distribution. Hoskins and Rodwell (1995) investigated the calculated response of the atmosphere, given the longitudinal average of the observed atmosphere as a background state, to heating over South Asia. In both upper and lower troposphere, the induced circulation encompasses about one-third of the globe longitudinally, and reaches across the equator and far into the southern hemisphere. Over the Indian Ocean, the dominant pattern of south-easterly flow south of the equator, crossing the equator over East Africa and becoming the south-westerly flow onto the coast of southern Asia, is well reproduced by the calculation, as is the large, zonally elongated, upper tropospheric anticyclone stretching from the coast of South-east Asia all the way to West Africa. There is a second anticyclone south of the equator, rather like a weak mirror image of that in the northern hemisphere, again in agreement with observations. Thus, most of the observed large-scale features of the tropical atmospheric circulation in northern summer, at least in the region from the Greenwich meridian to the east coast of Asia, can be understood as a response to the diabatic heating over South and South-east Asia. This underscores the far-reaching impact of the Indian Ocean monsoon system, not just to South and South-east Asia, but to the global atmosphere. In fact, Rodwell and Hoskins (1995; 2001) pointed out that one aspect of the calculated response is subsidence over northern subtropical Africa, i.e., the Sahara Desert region; hence, since subsidence is characterized by warm, dry air, there may be a direct link between desertification of the Sahara and the Asian monsoon. As noted above, the global desert belts are products of large-scale subsidence, itself a characteristic of the entire tropical circulation; the point here is that the reason for the size and extreme dryness of the Sahara is probably a result of its geographical location, relative to the Asian monsoon.

The circulation associated with the Asian monsoon, then, is quite well under-stood once the rainfall pattern is established. Moreover, the calculated low-level flow is such as to transport warm, moist air from over the warm waters of the North-west Indian Ocean onto the continent, thus supplying moisture to sustain the rainfall, and providing air of relatively high moist entropy that will be further heated over the land. Thus, the solution is qualitatively self-consistent. However, since the location of the rainfall is imposed in the cal-culation, the deeper issue of why the rainfall is located where it is remains unresolved. Why does the monsoon rainfall not extend, e.g., further west or further north, deep into the continent? Simple modeling studies (Chou *et al.*,

2001; Chou and Neelin, 2001; Privé and Plumb, 2007a,b) indicate that the bulk of deep moist convection, and correspondingly the core of the updraft region, must coincide with maximum boundary layer h_m.

In the Indian Ocean region, the envisaged evolution is therefore as follows. At the start of summer, maximum h_m is invariably found over the ocean, south of the equator; moist convection and rainfall maximize in the ITCZ, more-or-less coincident with the warmest SSTs (Figure 1.11). As summer progresses, the land starts to warm – as do the SSTs in the northern part of the ocean, especially in the Arabian Sea and the Bay of Bengal – and eventually h_m in the vicinity of the south coast of the continent exceeds that over the near-equatorial ocean, at which stage, almost simultaneously, the atmospheric circulation dramatically alters (very rapidly, as was seen in Figure 1.14) and intense rainfall sets in near and over the southern part of the continent. Thus, onset of the monsoon circulation is, as long recognized, associated with a reversal in the land–sea contrast, and that statement can now be made more quantitative: it is the contrast in the geographical distribution of boundary layer moist static energy that is the controlling factor. At first sight, it might appear that this distribution can be explained as a simple and direct consequence of SST over the ocean, and of solar energy input over the land, and that all the essential details are thus understood.

Inevitably, of course, it is not quite so simple. If it were, location of maximum h_m would, once moved onto the land, follow the Sun to $23°$ latitude whereas, in reality, the core of the monsoon rainfall is never found so far from the equator, especially in South Asia where, as we saw in Figure 1.16, much of the rain remains over the coastal ocean, while the inland penetration of monsoon rainfall into northern Australia is extremely limited, despite the vigorous cross-equatorial flow that supports it (Figure 1.7). To some extent, this may be a reminder that ocean SSTs near the coast themselves become very warm in early summer (although they cool in midsummer, a point to be raised in the following text). However, one must also recognize that the energy budgets over the land are not as simple as the foregoing discussion would imply. Once convection sets in over the land, clouds become a major, and complex, factor in the surface radiation budget. Moreover, radiative heating of the surface is not the only important factor: boundary layer air is also influenced by advection from elsewhere. Just how important this is, and the details of the key factors, depend on geographical details but, in general, air over the midlatitude continents is, even in summer, either dry (in locations uninfluenced by advection from the ocean) or cool (where advection of air from the relatively cool midlatitude ocean is important) and thus, either way, has relatively low h_m. If the monsoon circulation were to penetrate far enough inland to entrain such air, therefore, it would become progressively more difficult to sustain the high h_m required to drive the circulation in the first place. Thus, the low h_m of deep continental air may be the main factor limiting inland penetration (Chou and Neelin, 2001; Neelin, 2007). Moreover, this factor may at the same time limit the intensity of the circulation: a stronger low-level onshore flow would tend to move the convecting region further inland, thus reducing its maximum h_m, and hence reducing its intensity (Privé and Plumb, 2007b).

The potential limiting effects of the onshore flow remind us that the monsoon circulation is fundamentally dependent on surface conditions over both land and ocean. Moreover, sea-surface temperatures in the Indian Ocean are largely controlled by the surface winds, implying a feedback loop involving land, ocean and atmosphere. For one thing, the strong low-level monsoon winds over the Indian Ocean, and especially, over the Arabian Sea in northern summer induce strong sensible and evaporative heat loss to the atmosphere; this is one factor moderating SSTs in summer. In addition, surface temperatures over much of the Indian Ocean are strongly influenced by heat transport within the ocean (e.g., Godfrey et al., 1995), and the ocean circulation that effects this transport is itself wind-driven (McCreary et al., 1993). This has led to suggestions that the biennial oscillation of the Asia–Australia–Indian Ocean monsoon system is a manifestation of land–ocean–atmosphere interactions (Nicholls, 1983; Meehl, 1994, 1997). Meehl (1997) proposed a mechanism whereby the vigorous low-level winds in a strong monsoon year cool the surface waters of the equatorial Indian Ocean, leading to a weak monsoon the following year. Webster et al. (1998, 2002) incorporated ocean dynamics into Meehl's picture; the slow timescales of the ocean circulation provide the year-to-year memory that is needed to close the theory.

1.4.2   Role of orography

It has long been recognized, and explicitly demonstrated in general circulation models (Hahn and Manabe, 1975), that orographic effects are crucial in making the Asian monsoon the strongest on the planet. Several subsequent studies, using a wide range of such models, have confirmed the importance of Tibetan uplift to the present day monsoon, both in South and East Asia (Prell and Kutzbach, 1992; An et al., 2001; Liu and Yin, 2002; Abe et al., 2003; Kitoh, 2004). While Hahn and Manabe noted that orography can influence the monsoon through both mechanical and thermal effects, much emphasis in the subsequent literature was given to the latter, frequently through vague reference to elevated heating whose significance was unclear until Molnar and Emanuel (1999) demonstrated that, for an atmosphere in radiative–convective equilibrium, boundary layer moist static energy would be greatest, for the same solar input, over the highest elevations. Mechanically, mountains have direct effects, a local impact on precipitation through induced uplift, and a more widespread

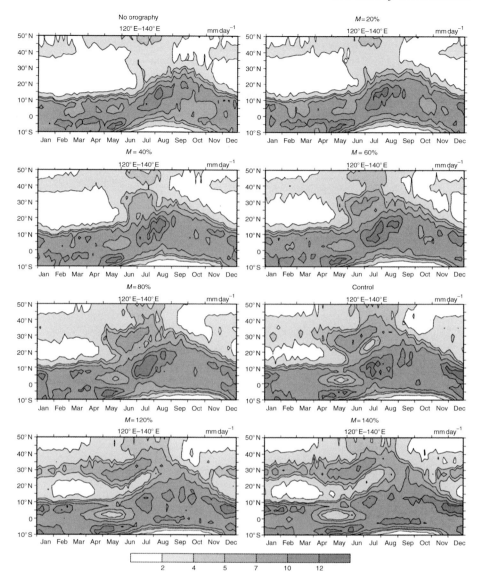

Figure 1.19 Average precipitation rate (mm day^{-1}) between 120° and 140° E from a general circulation model, in which present-day topographic height has been multiplied by a factor M, where M ranges from 0 through 140%, as indicated. (Kitoh, 2004.)

impact by shielding the monsoon region from the moderating influence of low-level air from higher latitudes with low moist static energy (Neelin, 2007; Privé and Plumb, 2007b). The former effect is visible in the rainfall distribution of Figure 1.16, while the latter may be important in North America as well as in Asia. A further effect, less direct but potentially of great importance, is on sea-surface temperature distributions via orographic influences on low-level winds (Kitoh, 1997; 2004); such influences can extend far beyond the location of the orography.

An example illustrating the dependence of modeled rainfall on orographic heights is shown in Figure 1.19. The results shown are of precipitation averaged over 120–140° E from a series of simulations by Kitoh (2004) using a coupled ocean–atmosphere general circulation model, with all surface elevations equal to present day values multiplied by a factor M. With $M = 0$, there is little evidence of a monsoon, most precipitation across the region remaining near the equator. The characteristic northward migration of precipitation in early summer first appears when $M = 40\%$; it strengthens as M is increased further. Only when orography is close to present values does heavy precipitation extend deep inland.

2

Controls on the Asian monsoon over tectonic timescales

2.1   Introduction

As described in Chapter 1 the intensity of the modern Asian monsoon can be understood as being principally a product of the seasonal temperature differences between the Indian and Pacific Oceans and the Asian continent. It is these differences that drive wind and weather systems in the strongly seasonal fashion that characterizes the climate throughout South and East Asia (Webster *et al.*, 1998; Figure 2.1). As a result, the long-term history of the Asian monsoon would be expected to extend as far back in time as the assembly of these major geographic features themselves, i.e., to the construction of Asia as we know it, following collision of the Indian and Asian continental blocks, generally presumed to be around 45 to 50 Ma (e.g., Rowley, 1996). Even the youngest estimates place the final collision of India and Asia at no younger than 35 Ma (Ali and Aitchison, 2005; Aitchison *et al.*, 2007).

Monsoon weather systems are not unique to Cenozoic Asia. Monsoon climate systems have been recognized in ancient supercontinents, such as Pangaea, where similar major temperature differences existed between land and ocean (e.g., Loope *et al.*, 2001). Indeed, even before the collision of India and Asia there must presumably have been a monsoon system operating between the Tethys Ocean in the south and Eurasia in the north. However, the intensity of that system would have been low because prior to the Indian collision Asia was somewhat smaller and less elevated than it has been for most of the Cenozoic (Wang, 2004). Furthermore, because global sea levels were higher in the early Cenozoic (Haq *et al.*, 1987) wide, shallow seas (Paratethys) covered much of the area of what is now central Asia. These waters had the effect of reducing the temperature contrast further and inhibiting monsoon strength (Ramstein *et al.*, 1997;

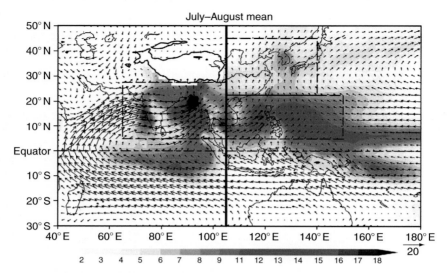

Figure 2.1 Climatological July–August mean precipitation rates (shading in mm day^{-1}) and 925 hPa wind vectors (arrows) from Wang *et al.* (2003a). The precipitation and wind climatology are derived from CMAP (Xie and Arkin, 1997) (1979–2000) and NCEP/NCAR reanalysis (1951–2000), respectively. The three boxes define major summer precipitation areas of the Indian tropical monsoon (5–27.5° N, 65–105° E), WNP tropical monsoon (5–22.5° N, 105–150° E) and the East Asian subtropical monsoon (22.5–45° N, 10–140° E). Reprinted with permission of Elsevier B.V. See color plate section.

Fluteau *et al.*, 1999). The modern summer monsoon is driven by anomalous heating of air over central Asia and the formation of a low pressure system that draws moist air from the surrounding oceans toward the continent resulting in the rains that characterize the system in South Asia and Indochina (Figure 2.2). The conditions responsible for the intensity of this low-pressure area in part reflect the solid Earth tectonic processes that have built modern Asia.

The intensity of both the winter and the summer monsoons has been strongly influenced in the recent geologic past by global climate changes driven by variability in the Earth's orbit over timescales of tens of thousands of years (e.g., Clemens and Prell, 2007), as discussed in detail in Chapter 4. The 20, 40 and 100 ky periods of the Milankovich cyclicity are clearly important in controlling the monsoon, especially since the initiation of Northern Hemispheric Glaciation around 2.7 Ma (Clemens and Prell, 1990). However, these fluctuations are superimposed on a longer-term evolution that is linked to the tectonic evolution of the solid Earth, principally in continental Asia and the Indian Ocean. In this chapter we explore the competing hypotheses that attempt to explain the intensity of the Asian monsoon on tectonic timescales (i.e., $>10^6$ y).

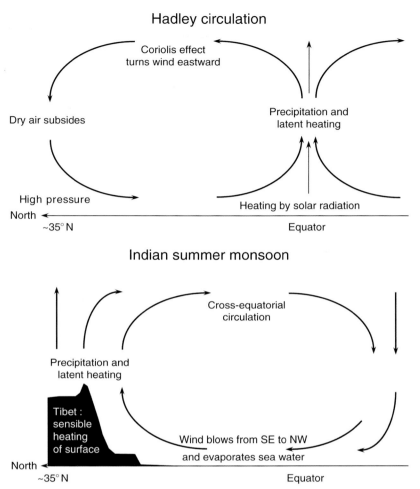

Figure 2.2 Cartoon from Molnar *et al.* (1993) illustrating the effect of the Tibetan Plateau on circulation over the Indian subcontinent and neighboring Arabian Sea. The top panel shows classic Hadley circulation, in which heating at the equator evaporates sea water, thereby reducing the mean density of the air causing it to rise. Condensation and precipitation leave latent heat of vaporization for additional heating, causing air to ascend further. The lower panel shows Indian summer monsoon circulation. With a strong heat source over Tibet in Summer, air rises near and above the plateau and then flows toward the equator. Air drawn from equatorial regions into the ascending limb of the cell near and above the plateau produces the monsoonal winds that characterize the Indian summer monsoon. Reprinted with permission of the American Geophysical Union.

2.2   The influence of Tibet

Tibet, the Himalayas and the many ranges of mountains that lie largely north and east of the main plateau (Figure 2.3) represent the largest mountain chain on Earth since the Pan-African Orogeny, approximately 900 million years ago. As such it is not surprising that this broad edifice has affected the global climate system in a number of ways, but especially with regard to the Asian Monsoon. Tibet has also been implicated in the onset of Northern Hemispheric Glaciation since ~2.7 Ma, as a result of intense chemical weathering, mostly of Himalayan rocks. Chemical reactions that occur during alteration of silicate rocks consume CO_2, resulting in a reduction in global atmospheric CO_2 levels. Because CO_2 is a greenhouse gas the ongoing chemical weathering of silicate rocks exposed in the Himalaya, especially since the intensification of the monsoon, has been an important global climatic influence (e.g., Raymo et al., 1988; Raymo and Ruddiman, 1992).

Prell and Kutzbach (1992) were the first to make an explicit link between the modern elevation of the Tibetan Plateau and the strength of the monsoon. Although this was an important new development, the ability of broad, elevated regions to cause disruption to atmospheric circulation, and thus to influence local and global climate had been recognized before that time. For example, the elevated regions around the Colorado Plateau in western North America had been singled out as being a possible influence on northern hemispheric climate (e.g., Kutzbach et al., 1989; Ruddiman and Kutzbach, 1989). In both central Asia and western North America the key factor in the ability of mountains to cause climate change is not only their elevation but also their lateral extent (Figure 2.4). While high but narrow mountain ranges generate orographic rainfall along the range front, their ability to influence atmospheric circulation patterns on a regional scale is rather restricted. However, it is recognized that because of their great size the plateaus of Tibet and western North America do exert important influences on regional and even global climate because of their ability to disrupt atmospheric circulation patterns (e.g., the path of the Jet Stream), as well as to generate summer low-pressure systems that draw in moist air from the surrounding oceans, while reversing the process during the winter months.

Numerical climate modeling by Prell and Kutzbach (1992), as well as, subsequently, by Kutzbach et al. (1993), An et al. (2001) and Kitoh (2004), emphasized the need for Tibet to be not only high (at least 3 km) but also broad before it begins to influence the climate system. Kitoh (2004) estimated that rainfall would not be greatly elevated in the central Himalayas until around 40% of the present topography was generated (Figure 2.5). Because Tibet is unique on the modern planet surface in terms of its altitude and extent it exercises a

Figure 2.3 Shaded topographic map of South and East Asia showing the dominant high topography of the Tibetan Plateau and the associated elevated plateaus and ranges that make up the widest and highest modern orogenic region on Earth. Labels show the locations of the major geographic features named in the text.

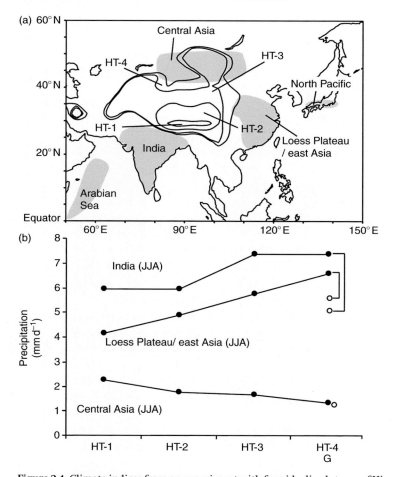

Figure 2.4 Climate indices from an experiment with four idealized stages of Himalaya–Tibetan plateau elevation (HT-1 to HT-4) and one glacial maximum stage (G) made with the NCAR climate model CCM3. (a) Areas (in gray) for which climate indices are summarized (below), and approximate boundaries of the idealized topography stages with elevations higher than 1000 m outlined: HT-1, small elevated region, with maximum elevation less than 1700 m; HT-2, Himalaya and Tibetan Plateau of limited north-south and east-west extent with maximum elevation 2700 m; HT-3, Himalaya and Tibetan Plateau considerably extended to the north and west with maximum elevation 5700 m; and HT-4, modern, with extension of the plateau along the eastern and northern margins and maximum elevation 5700 m. The elevations used in the climate model reflect a smoothing of the topography consistent with the spatial resolution of the climate model, and are significantly lower than the observed or estimated elevations. (b) The June–July–August (JJA) precipitation for India, the Loess Plateau and east Asia, and central Asia, for four simulations (HT-1 to HT-4) with progressive increase in mountain–plateau elevation and one simulation (G) with glacial-age modifications to HT-4 (lowered atmospheric CO_2 concentration, enlarged Northern Hemisphere ice sheets, lowered sea-surface temperatures). The climate values for G are indicated with an open circle connected to the climate value for HT-4 by a thin vertical line. This figure is from An *et al.* (2001), reproduced with permission of Macmillan Magazines Ltd.

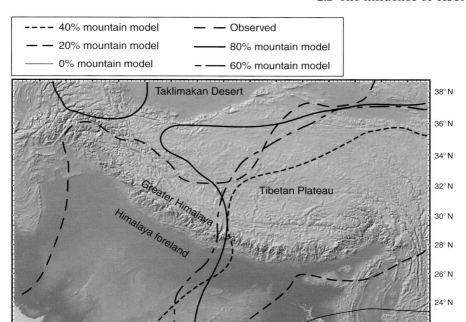

Figure 2.5 Map showing the extent of the $1.0\,\text{mm}\,\text{day}^{-1}$ mean precipitation for June observed and for a series of climate models predicting the intensity of summer rains for Himalayan–Tibetan topography at different percentages of its current elevation (Kitoh, 2004). Note that the rain front only migrates significantly inland after 40% plateau elevation has been achieved.

disproportionately strong influence on global climate. Research is continuing to determine the detailed relationship between Tibetan topography and monsoon intensity. For example, which areas of Tibetan elevation are crucial to controlling certain aspects of the Asian monsoon? Chakraborty *et al.* (2002) argued that the high ranges of the Karakoram, western Himalaya and western Tibet are the primary geological influence on the strength of the African and SW Asian monsoon, as opposed to the East Asian monsoon, which instead would be influenced more by the eastern plateau and the temperature of the western Pacific. Chakraborty *et al.* (2002) also noted the potential for topography in East Africa to modify the intensity of the SW Asian monsoon. Figure 2.6 shows their model, which predicts that while the intensity of the summer monsoon winds in the Arabian Sea is relatively unaffected by the lack of Tibetan high topography east of 80° E, the same is not true when the western plateau is also eliminated from the simulation. In either case, what is not yet clear is when Tibet breached the crucial threshold that allowed the monsoon to intensify.

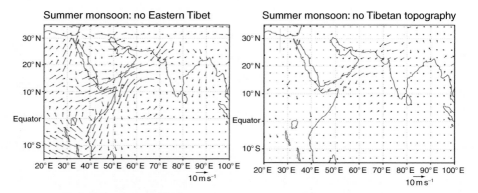

Figure 2.6 Left: Difference in horizontal wind vector between a climate model run with no global topography and a control run for July 1998 (Chakraborty *et al.*, 2002). Right: Difference in horizontal wind vector between a climate model run with no Tibetan Plateau east of 80° E and a control run for July 1998 (Chakraborty *et al.*, 2002). The model suggests that the eastern plateau has little influence on the SW summer monsoon. Reprinted with permission of American Geophysical Union.

2.2.1   *Tectonic evidence for Tibetan uplift*

Reconstructing the surface uplift history of Tibet in order to compare it with monsoon climate records has been a major goal for tectonic geologists for several years. Tibet's altitude is important to geodynamic models of strain accommodation in continental collision zones, as well as to climate–tectonic coupling hypotheses. The task is complex because the record of elevation preserved on land is hard to interpret because the terrestrial sedimentary record is patchy and often difficult to date accurately. Continental sediments are largely free of the marine fossils that typically provide age control to sedimentary geologists. Part of the lack of consensus on when Tibet experienced major regional surface uplift derives from different scientists using data from different, and limited, parts of Tibet to draw conclusions about the entire plateau. In contrast to the Himalayas, which grew from below sea level, the southern edge of Eurasia appears to have been elevated as an Andean-type subduction margin before the collision (e.g., England and Searle, 1986; Murphy *et al.*, 1997) and almost surely was not a lowland. Recognition of thrust faulting and folding of Cretaceous sedimentary rocks, together with early Cenozoic angular unconformities above the deformed Mesozoic strata, requires extensive pre-collisional crustal shortening (Burg *et al.*, 1983). Whether or not southern Tibet had attained a pre-collisional height of 3–4 km, as Murphy *et al.* (1997) have argued, remains a question not easily addressed with existing data. In any case, the growth of the Tibetan Plateau to a great height over a large area must have occurred since the start of India–Asia collision around 45–50 Ma.

One method employed for constraining Tibetan uplift is to date the N–S oriented normal faults that crosscut much of the plateau. This approach is based on the hypothesis that the current E–W extension of the plateau is caused by Tibet's gravitational collapse after reaching a maximum altitude in the geologic past (England and Houseman, 1989). Thus, the oldest extensional faults could potentially record the time at which maximum altitude was attained. Harrison *et al.* (1992) and Pan and Kidd (1992) used such logic and radiometric ages from rocks exposed along extensional faults, especially in the vicinity of Lhasa, in the Nyainqentanglha Range in southern Tibet, to argue for 8 Ma being the time of maximum elevation in Tibet. However, in other parts of southern Tibet faulting and rock exhumation are dated as accelerating in Early Miocene time, near 20 Ma (e.g., Copeland *et al.*, 1987; Harrison *et al.*, 1992), although this appears to be at least in part related to Himalayan N–S extension rather than the stress in the wider Tibetan Plateau.

The hypothesis that 8 Ma was a tectonically crucial time in Tibet was further strengthened by the seismic and geophysical imaging of a series of long-wavelength buckles in the Indian Ocean lithosphere south of Sri Lanka. These are atypical of the strong oceanic lithosphere seen elsewhere (e.g., Bull and Scrutton, 1990; Cochran, 1990; Figure 2.7). Subsequent drilling of the sediments overlying the folds and faults at Ocean Drilling Program Sites 717–719 (Figure 2.8) demonstrated that the most important phase of folding occurred around 8 Ma (Cochran, 1990; Krishna *et al.*, 2001), indicating a regional maximum in tectonic stress. It was argued that this is consistent with Tibet rapidly uplifting prior to that time (Molnar *et al.*, 1993).

Although detailed discussion of the tectonic mechanisms by which Tibet became elevated lies outside the scope of this book, and is provided in the recent review by Harris (2006), it is worth briefly mentioning one of the more favored type of models, typified by the work of Bird (1978), Molnar *et al.* (1993) and Fielding (1996). In the view of these models the crust and mantle lithosphere under Tibet thickened progressively owing to horizontal compression after India began to collide with Asia. Although the thicker crust resulted in surface uplift owing to simple isostatic balancing of the buoyant crust, Molnar *et al.* (1993) calculated that for the approximate doubling of the crustal thickness observed an altitude of 4 km would be predicted, somewhat short of the 5.5 km typical over much of the central plateau today. An additional rapid increase in surface elevation was predicted to occur when the dense mantle part of the lithospheric root under Tibet became too thick and gravitationally unstable. Its catastrophic or more gradual convective loss back into the asthenosphere (Houseman *et al.*, 1981) would have released a heavy anchor from the base of the plateau, allowing the whole edifice to rise up rapidly at that time (Figure 2.9).

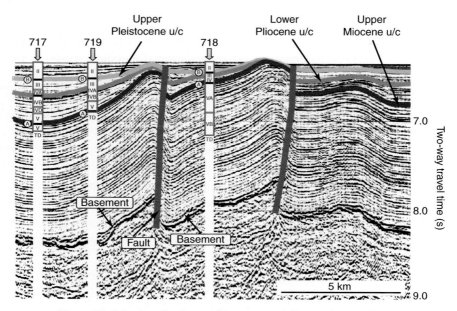

Figure 2.7 Seismic reflection profile across anticlines and thrust faults in the Indian Ocean showing the nonconformity of Horizon A with underlying layers that have been warped over folds and thrust faults in the basement. Numbers at the top mark drilling sites of Ocean Drilling Program Leg 116 (modified from Bull and Scrutton, 1990). Reproduced with permission of Macmillan Magazines Ltd.

Attempts to identify this delamination (and uplift) event have involved dating the plateau-wide potassic magmatism (e.g., Turner *et al.*, 1993) that is proposed to be linked to the loss of the lithospheric keel. However, age dates for these volcanic rocks range from 25 and 19 Ma from southern to northern Tibet, a long time before the proposed onset of major extensional faulting at 8 Ma. Chung *et al.* (1998) also noted that potassic volcanic rocks are older in eastern Tibet, dating back to 40 Ma, and younger in the west, though it is not clear how that pattern relates to mantle lithospheric loss and surface uplift. The significance of volcanism to surface uplift is further complicated by the studies of Ding *et al.* (2003) and Williams *et al.* (2004) who have argued that the distribution and chemistry of the volcanism is better explained by continued ongoing continental subduction within the plateau, and has nothing to do with surface uplift.

Belief in a relatively simple pattern of rapid plateau-wide surface uplift, followed by gravitational collapse at 8 Ma was further undermined by additional dating studies of extensional faulting from other parts of the plateau away from the Lhasa region (e.g., Coleman and Hodges, 1995; Blisniuk *et al.*, 2001; Williams *et al.*, 2001). These constraints showed that E–W extension had started

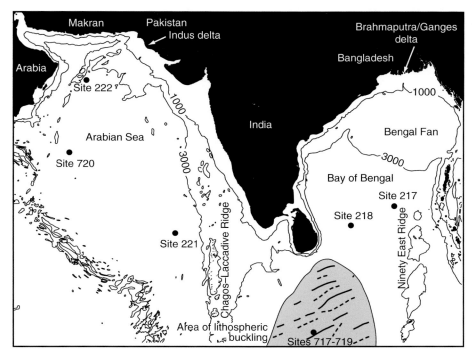

Figure 2.8 Simplified bathymetric map of the northern Indian Ocean showing the area of lithospheric buckling dated at 8 Ma by drilling at ODP Sites 717–719. Solid lines show the crests of anticlines, while dashed lines show traces of synclines mapped by Bull and Scrutton (1992). Water depths are shown in meters and are from GEBCO compilation. Deep Sea Drilling Project and Ocean Drilling Program sites penetrating the Indus and Bengal submarine fans are shown as filled circles.

as early as ~19 Ma, although again most of these data are from limited study areas, mostly in southern and central Tibet, and their application to the entire plateau is questionable.

In an alternative approach to tracing Tibetan elevation, other workers have used the age of compressional deformation to constrain the timing of surface uplift because this must reflect crustal thickening. These ages do not tell when the plateau reached altitudes that would have been important in triggering a monsoon climate, but they can at least set a maximum age for the start of uplift. Tapponnier *et al.* (2001), in a major synthesis, proposed that southern Tibet had already developed into an elevated plateau during the Eocene, with the age of deformation in each successive tectonic block younging gradually toward the north. In this scenario the north-eastern plateau around the Qaidam Basin dates from the Pliocene, as also suggested by a number of independent studies (e.g., George *et al.*, 2001; Garzione *et al.*, 2005). However, Fang *et al.* (2003) argue

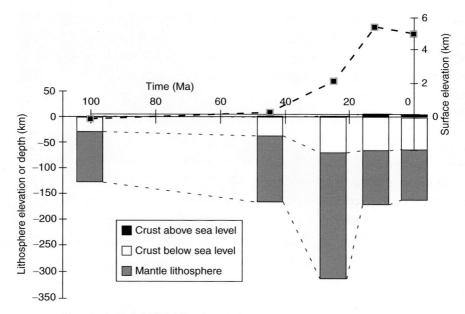

Figure 2.9 Model of Fielding (1996) showing isostatic uplift of Tibet at five approximate times (100, 45, 25, 8 and 0 Ma). The estimated lithospheric structure is shown as a column with the mantle lithosphere, crust below sea level and crust above sea level shaded according to the key. For the same time the elevation is also shown as black squares, vertically exaggerated according to the scale on the right. Reprinted with permission of Elsevier B.V.

that even this region had begun to compress and become elevated as early as the Oligocene (>24 Ma), even if major uplift postdated 8 Ma.

It is important to note that because Tibet is very large and wide it probably did not behave as a simple block throughout its development. Indeed, some geodynamic models that attempt to understand Tibet as being the product of deformation of a viscous fluid crust would predict that the plateau should grow progressively vertically, and especially laterally, as India continues to penetrate into Eurasia (Houseman and England, 1993; Clark and Royden, 2000). If this is correct then the idea of being able to generate a simple plateau-wide uplift history is erroneous and a more sophisticated approach is required.

2.2.2   *Altitude proxies*

Study of the timing of deformation in Tibet is important to constraining when monsoon-related surface uplift might have occurred, but it is not a direct measure of elevation. To actually quantify paleo-elevation the flora preserved in sediments deposited on the plateau have been the focus of study. Their use is based on the idea that as a given area of Tibet is uplifted the air at ground level must cool, thus changing the vegetation that can thrive there. Paleobotanical finds (pollen and fossil plant organs) have been used to imply a regional cooling of air temperatures over Tibet, and these have, in turn, been interpreted as evidence for a rise of the plateau at an accelerating rate since Eocene time (e.g., Axelrod, 1980; Mercier *et al.*, 1987; Powell, 1986; Xu, 1979; Zhao and Morgan, 1987). However, because most of these studies ignore global cooling over the same period (e.g., Zachos *et al.*, 2001), and because much of the evidence is based on drawing analogies with present-day plants without allowing for evolutionary develop-ment, such paleo-altitude estimates are suspect (e.g., Molnar and England, 1990).

Only recently have paleobotanists been able to correct for the long-term global cooling effect. Spicer *et al.* (2003) exploited the empirical relationship between leaf physiognomy and properties of the atmosphere related to altitude by using a computational method, known as the climate leaf analysis multivari-ate program (CLAMP; Wolfe *et al.*, 1998). Although empirical, this relationship seems to be robust throughout the Tertiary, in part because leaf physiognomy has a basis in the convergent evolution constrained by the laws of physics, and in part because the relationship between leaf morphology and climate is not subject to diagenetic modification. Spicer *et al.* (2003) used such evidence to propose that southern and central Tibet were close to their current altitude by at least ~15 Ma, earlier than proposed by Harrison *et al.* (1992) and Molnar *et al.* (1993), but consistent with the older ages for E–W extension (e.g., Blisniuk *et al.*, 2001).

Constraints on altitude have also been derived from the study of carbon isotopes in horse and rhino tooth fossil enamel found in Upper Miocene sedi-ments exposed in southern Tibet. The isotope character of the tooth material is dictated by the diet of the animals and the indication from these 7 Ma old records is that the flora was dominated by C4 grasses at that time (Wang *et al.*, 2006). Such plants are most typical of warmer climates and thus lower altitudes than seen at present. Even after accounting for late Cenozoic global cooling and paleo-atmospheric CO_2 levels, these data would suggest that this part of southern Tibet was less than 2900–3400 m above sea level at 7 Ma, and thus that signifi-cant uplift had occurred since that time. Quite how these data can be reconciled with the paleobotany work of Spicer *et al.* (2003) is not presently clear.

An independent line of evidence is, however, supplied by the analysis of the oxygen isotopic character of carbonate sediments from soils and lake sediments, largely from southern Tibet. This method exploits the fact that rainfall at higher elevations is more concentrated in isotopically light oxygen compared with low lands, i.e., $\delta^{18}O$ is systematically more negative at higher altitudes (Figure 2.10). This signature is then transferred to carbonates as they crystallize in lake sediments or in soils, such as calcretes. Garzione *et al.* (2000, 2004) and Rowley *et al.* (2001) have shown that southern Tibet must have been close to modern

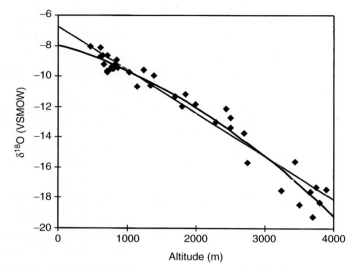

Figure 2.10 $\delta^{18}O$ vs. altitude for tributaries sampled by Garzione *et al.* (2000) along the Kali Gandaki and Tinau Khola in West–Central Nepal during September and October 1999. Only those tributaries with perennial flow over an elevation range of <2000 m are plotted in the arid region north of the high Himalayas. Line fit and second-order polynomial fit are shown as black lines. Reprinted with permission of Elsevier B.V.

elevations prior to 10 Ma, although when that altitude was first reached is not yet constrained by this method. Further north, Currie *et al.* (2005) showed that central Tibet was close to modern elevations prior to 15 Ma, extending the evidence for a wide, high plateau prior to the 8 Ma favored by earlier studies. Similarly, Dettman *et al.* (2003) showed a change in the oxygen isotope composition of lake carbonates at 12 Ma in NE Tibet, suggesting a change in atmospheric circulation patterns driven by uplift at that time. More recently, and most dramatically, Rowley and Currie (2006) used Upper Eocene sediments from the Lunpola Basin in the centre of the plateau to propose that that region has been close to modern elevations since at least 35 Ma, far earlier than previously suggested.

2.2.3   Erosion and the uplift of Tibet

The erosional record of Tibet preserved in the sedimentary basins that surround the plateau also provides some control on the age of Tibetan uplift. Erosional records must be treated with caution because they do not directly measure surface uplift. Nonetheless, the logic in this approach is based on the premise that as the plateau rises it increases the gradient of rivers around its periphery and drives faster erosion as rivers incise gorges in the flanks.

One potential complexity with this approach is that plateau uplift can only cause erosion if there is precipitation to incise the river valleys, meaning that lack of erosion does not preclude uplift.

Typical of this type of study was the identification of a dramatic influx of conglomerate to the Tarim Basin during the Pliocene (Zheng *et al.*, 2000). This was interpreted to be the product of major surface uplift, at least in Northern Tibet, rather than a response to a more erosive climate since the onset of Northern Hemispheric Glaciation (Zhang *et al.*, 2001). Wang (2004) has argued that Plio-Pleistocene uplift is even more widespread and is responsible for the generation of much of the modern drainage in East Asia and even for the Siberian rivers draining into the Arctic Ocean. In particular, Wang (2004) suggested that the Yangtze (Changjiang) River is only Plio-Pleistocene in age based on core and geophysical data from the vicinity of the modern delta (Li *et al.*, 2000) and is formed as a direct result of major Tibetan surface uplift in the last few million years.

The development and erosion of rivers caused by Tibetan uplift and monsoon intensification is discussed further in Chapter 5, yet it should be noted that the concept of plateau uplift driving re-organization of drainage is now well accepted (Brookfield, 1998; Clark *et al.*, 2004). In this scenario, rock and surface uplift causes re-tilting of continental topographic gradients over long wavelengths, driving drainage capture and transfer of headwaters from one system to another. It is hard to determine when Asian rivers started to re-organize into their present nondendritic patterns because river sediments are often poorly preserved onshore over long periods of geologic time. Analysis of sediment volumes and isotopic compositions from the Red River delta region now indicate that, in contrast to the Yangtze, this river at least had experienced major drainage capture before 24 Ma, implying the start of Tibetan surface uplift during the Oligocene, even in this south-eastern edge of the plateau (Clift *et al.*, 2006a).

Very young Plio-Pleistocene ages for the onset of major surface uplift are also hard to reconcile with the altitude proxies discussed above and with new exhumation data from the gorges of the Yangtze system in Yunnan. Clark *et al.* (2005) used the (U-Th)/He low-temperature thermochronology method for charting the cooling of rocks in the upper few kilometers of the crust in order to date the incision of gorges in that part of Tibet. This method allows cooling of apatite crystals through the 40–80 °C temperature range to be dated (Farley, 2000), which in most continental areas corresponds to a depth of around 1.3–2.6 km. These workers used this information to suggest incision and thus plateau surface uplift (at least in that part of Yunnan) at 13–9 Ma. These data do not preclude older plateau uplift further west, but they do argue against the onset of major plateau-wide uplift during the Plio-Pleistocene.

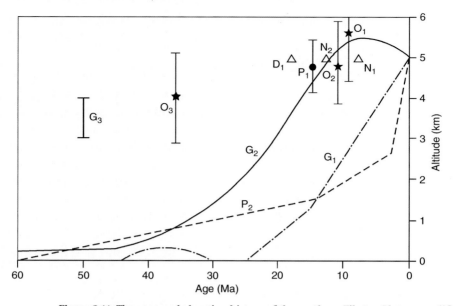

Figure 2.11 The proposed elevation history of the southern Tibetan Plateau modified after Harris (2006). Sources of paleo-altitude studies as follows. Normal faults; N_1 = Harrison *et al.* (1995), Yin *et al.* (1999), N_2 = Coleman and Hodges (1995). Dykes; D_1 = Williams *et al.* (2001). Oxygen isotope studies; O_1 = Rowley *et al.* (2001), O_2 = Garzione *et al.* (2000), O_3 = Rowley and Currie (2006). Paleobotany sites; P_1 = Spicer *et al.* (2003), P_2 = Xu (1979). Geodynamic models; G_1 = Zhao and Morgan (1987), G_2 = Fielding (1996), G_3 = Murphy *et al.* (1997). Reprinted with permission of Elsevier B.V.

It is clear that the proposed age of plateau uplift varies wildly owing to the lack of a complete, unambiguous record of altitude spread across Tibet. Because the timing of Tibetan uplift is still strongly debated, determining the relationships between tectonics, climate and erosion is difficult at present. Figure 2.11 shows the results of a recent compilation by Harris (2006), comparing a variety of competing Tibetan uplift models and the limited number of reliable data points for altitude. While the debate continues, it is becoming clearer that southern and maybe central Tibet was probably elevated close to modern values before 15 Ma and that older models invoking the bulk of plateau uplift occurring close to 8 Ma are likely invalid, even though this was clearly a time of growth along the NE and SE flanks (Clark *et al.*, 2005; Schoenbohm *et al.*, 2006). The older Paleogene history of Tibet, however, remains enigmatic, and the resolution of many paleo-altimeters is often insufficient to rule out an increase of less than 1 km after 8 Ma, owing to lithospheric delamination.

2.3   Oceanic controls on monsoon intensity

Solid Earth tectonic processes have influenced the intensity of the Asian monsoon over long timescales in other ways, as well as through topographic uplift of Tibet. Changing the temperature differential between Asia and the Indian Ocean may also be achieved through changes in the water masses in the Indian Ocean, which are also open to tectonic control. Closure of the Indonesian gateway between the Pacific and the Indian Oceans was one of the most important events to affect the oceanography and climate in South Asia. In the modern oceans the flow of water from the Pacific to the Indian Ocean is restricted by the island chains of Indonesia (Gordon and Fine, 1996). The waters that do flow through the various gateways are typically from the North Pacific and represent an important component of the Indian Ocean water mass (Godfrey and Golding, 1981; Tomczak and Godfrey, 1994; Fine et al., 1994). Because Australia has been moving north through the Cenozoic and has only come into collision with Indonesia during the Early Miocene (Hall, 2002) it follows that this flow into the Indian Ocean must have been stronger in the past. Field and seismic stratigraphic studies of the sediments in the Indonesian collision zone indicate that the collision process was advanced by the Early Pliocene (e.g., Pairault et al., 2003), but may have been initiated much earlier, at 16–18 Ma (Rutherford et al., 2001). The issue is controversial and many authors favor a younger age for initial collision that cut first deep-water, then shallow-water interchange between the two basins after around 12–13 Ma (Audley-Charles, 2004; Nathan et al., 2003; Nathan and Leckie, 2004). However, the most recent revision of the paleogeography based on sedimentary facies and paleomagnetic data lead Kuhnt et al. (2004) to propose loss of deep-water flow as early as about 25 Ma (latest Oligocene).

The collision and severance of the Indonesian through-flow has clearly been gradual and ongoing. Cane and Molnar (2001) modeled the flux of water through the Indonesian gateway and noted that the northward drift of Australia and New Guinea has resulted in a termination of flow from the warmer South Pacific around 3 Ma. This partial gateway closure reduced total flux and also diverted waters from the North Pacific into the Indian Ocean. North Pacific water is less saline and colder than its southern equivalent and by passing on some of these properties to the Indian Ocean this switch has acted to cool the Indian Ocean and reduce the intensity of the South Asian summer monsoons. Because the flow-through waters move westwards toward Africa after they enter the Indian Ocean (Godfrey, 1996; Gordon and Fine, 1996) this progressive closure seems to have had most effect on the African monsoonal climate, leading to increased aridity in eastern Africa after 3 Ma. However, climate modeling by Cane and Molnar (2001) also predicted moderate temperature decreases even

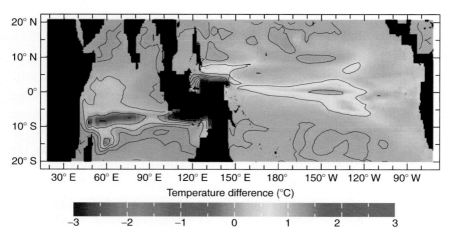

Figure 2.12 The difference in temperatures at 100 m depth between two ocean GCM runs from Cane and Molnar (2001). The sole difference between conditions for the runs is that in one the northern tip of New Guinea–Halmahera is at 28° N, and in the other it is at 38° S. Cane and Molnar suggest that the pattern showing the difference of the former minus the latter approximates the difference between that at present minus that at 4 Ma. Reproduced with permission of Macmillan Magazines Ltd. See color plate section.

in the Arabian Sea (Figure 2.12), owing to the change in flow through, which in turn reduces the intensity of the summer SW monsoon.

Because of the teleconnections between Asian and Atlantic climate, discussed more fully in Chapter 4, monsoon strength is also presumed to be affected by the closure of the Panama gateway between Atlantic and Pacific that may have initiated the modern North Atlantic Circulation and thus widespread Northern Hemispheric Glaciation after 2.7 Ma (Haug and Tiedemann, 1998). Gateway closure in Panama at around 4.6 Ma might be expected to affect the monsoon strength indirectly through changing global climate, but clearly has no direct impact on Indian Ocean water temperatures.

The earlier presence of shallow seas north and west of modern Tibet may be another important influence on monsoon strength. Ramstein *et al.* (1997) and Fluteau *et al.* (1999) used climatic modeling techniques to predict that monsoonal strength would have been influenced on geological timescales by the retreat of shallow seas from what is now Central Asia. In this model the epicontinental seas that flooded much of central Asia during the Paleogene progressively contracted, driven by falling global sea level and the uplift of the continental crust around Tibet and Mongolia. Using published paleogeographic reconstructions of Asia at 30 and 10 Ma (Dercourt *et al.*, 1993) these studies predicted monsoon strength in the context of the retreating seas and growing Tibet (Figure 2.13). The subaerial exposure of more continental crust in central

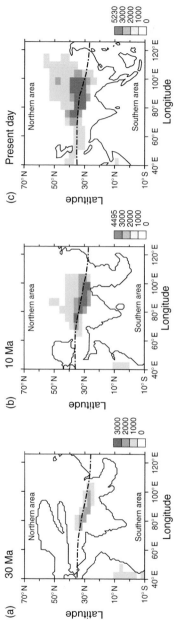

Figure 2.13 Eurasian paleogeography and topography after Ramstein *et al.* (1997). (a) Early Oligocene, (b) Late Miocene, (c) present day. The shaded areas represent the relief interpolated into the coarse grid of the model. Shading indicates average elevation above sea level. Progressive isolation of the intracontinental Parathethys Sea is caused by the topographic uplift and infilling of the seas by erosional flux from the collisional mountains. Reproduced with permission of Macmillan Magazines Ltd.

47

Asia affected air circulation over the region because it allowed the atmosphere to heat more under the influence of the summer sun, driven by the lower albedo of land compared with ocean waters. As a result the Central Asian summer low-pressure system is predicted to have become deeper and the monsoon stronger as the Paratethys shrank. The last direct marine connection to the Indian Ocean was cut around 18 Ma (Rögl and Steininger, 1984), making the thermal isolation of Asia from the Indian Ocean more intense. While this very early retreat of seas is more difficult to correlate with reconstructions that emphasize Late Miocene–Pliocene monsoon strengthening, this model may yet prove important as our understanding of the paleo-monsoon improves.

2.4   Summary

Determining whether the proposed connections between Tibetan surface uplift, Indonesian gateway closure or retreat of epicontinental seas and the intensification of the Asian monsoon are correct or not requires independent reconstructions of how each of these processes have evolved. As explained above, our understanding of Tibetan elevation is making progress, but continues to be a controversial subject. Although rock and surface uplift are clearly still active, modern paleo-altitude studies are now providing an image of southern and central Tibet being close to modern elevations certainly no later than the Middle Miocene (~15 Ma), and potentially much earlier. Data from south-east and northern Tibet point to more recent uplift in these peripheries. Theoretical models support field data for a progressive increase in the plateau's extent from a southern origin since the Indian collision.

Furthermore, it has historically been problematic to find unambiguous proxies for monsoon intensity in the geological record, making clear correlation difficult. In Chapter 3 we explore the different ways in which ocean and Earth scientists have attempted to track monsoon development and what has been learnt regarding the evolution of the phenomenon over tectonic timescales (i.e., >1 My).

3

Monsoon evolution on tectonic timescales

3.1   Proxies for monsoon intensity

If models that propose coupling between solid Earth tectonic processes and monsoon strength are to be tested then we must first produce reconstructions of monsoon evolution over long, tectonic timescales that can be compared with histories of mountain building or ocean closure. In this chapter we review the existing evidence for such development, typically over spans longer than one million years. We examine evidence from both oceanic and continental sources to see how the oceanography and climatic evolution are linked and then comment on how these relate to the tectonic evolution of Asia discussed in Chapter 2. Finding proxies for "monsoon intensity" in the seas that surround Asia is a complicated task because the nature of the monsoon changes from the Arabian to the South China Seas, as well as going from south to north Asia. Furthermore, the winter monsoon is often manifest in a different fashion than the summer monsoon, requiring separate proxies to be developed for each part of the system. For example, while the monsoon in the Arabian Sea is strongest in the summer and is characterized by powerful winds and oceanic upwelling, the monsoon in Bangladesh is largely characterized by heavy summer rains. Conversely, in northern China, while the monsoon does involve summer rains, it is dominated in that area by winter winds and dust storms. Thus the prospect of a "silver bullet" monsoon proxy that can be applied to areas at all periods of geological time is not likely.

3.2   Monsoon reconstruction by oceanic upwelling

3.2.1   *Monsoon winds in the Arabian Sea*

What the monsoon in all areas has as a common property is enhanced wind strength. The problem comes in finding a unique proxy for wind velocity that can be preserved in the geological record. This is especially true when the wind reverses direction between seasons. The Arabian Sea was one of the first regions where monsoon wind strength was reconstructed over long time periods using the marine sedimentary record. This is possible because the wind blowing to the north-east out of Arabia toward the Indian subcontinent carries large volumes of dust picked up from the deserts (Figure 3.1(a)), much of which is blown out to sea during the peak periods of the summer monsoon (Figure 3.2). This south-west monsoon induces upwelling of deep, nutrient-rich waters along the coast of Oman because it drives surface water away from the coast of Arabia, allowing deep water to rise during the summer (Anderson *et al.*, 1992; Curry *et al.*, 1992). Conversely, in the winter, winds from the north push water away from the coast of Pakistan, driving upwelling there, especially along the Makran coast (Figure 3.1(b)). Both stronger dust transport and enhanced upwelling leave a record in the marine sediments deposited in the deep-water Arabian Sea, which can thus be used to chart ancient monsoon strength when cored and analyzed.

In the modern Arabian Sea sediment trap experiments show that certain types of planktonic foraminifers become very abundant during the months of summer monsoon-driven upwelling (Prell and Curry, 1981). In particular, *Globigerina bulloides* became recognized as a form whose numbers were closely linked to the intensity of upwelling because that species thrives in the presence of colder, nutrient-rich waters. As a result of the enormous flux of fluvial sediment

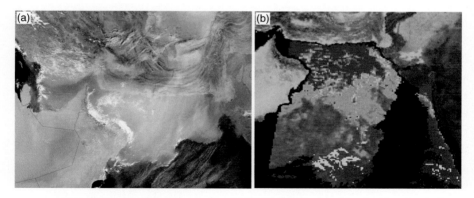

Figure 3.1 (a) A MODIS image from NASA (Visible Earth) of the Arabian Sea during the summer monsoon showing a storm blowing dust from Arabia to the north-east across the Gulf of Oman toward Pakistan. (b) A false color image from NASA showing chlorophyll in surface waters during the winter monsoon. Wind blowing south is driving upwelling and enhanced productivity along the Makran coast. Images reproduced with permission of NASA. See color plate section.

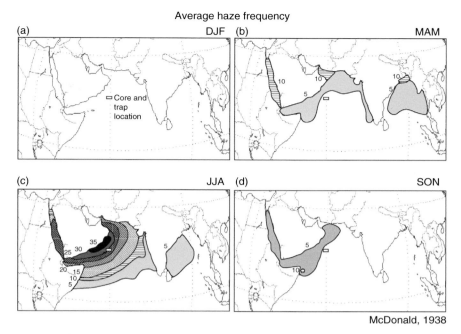

Figure 3.2 Map showing the changing haze at sea from Clemens and Prell (1990), based on data from McDonald (1938). (a) December–February, (b) March–May, (c) June–August, (d) September–November. Mapping clearly shows that the summer months are the peak time for dust transport out of Arabia across the Arabian Sea. Reprinted with permission of the American Geophysical Union.

reaching the ocean from the Indus Delta it is not practical to look at the relative abundance of *G. bulloides* or even wind-blown dust using cores from the deep sea sediment repository of that river, the Indus Fan, because there these signals are heavily diluted. Instead, coring on the continental margin of Oman was targeted by the Ocean Drilling Program (ODP) as a good place to assess monsoon strength because of the negligible run-off from rivers from Arabia, and also because this is the region of greatest summer monsoon-driven upwelling (Figure 3.3). Further scientific coring was undertaken on the Owen Ridge, which sits further offshore but since 20 Ma has been elevated above the level of the Indus Fan and is ideally located to record pelagic processes in the Arabian Sea (Mountain and Prell, 1990).

In a ground-breaking study Kroon *et al.* (1991) were able to apply the relationship linking *G. bulloides* abundance to summer monsoon wind strength to core samples taken from ODP Sites 722 and 727. Figure 3.4 shows the result of this work and specifically that the numbers of *G. bulloides* increased sharply after ~8 Ma. This observation provided the first strong evidence that the summer monsoon might have intensified at that time in the Arabian Sea. Although there

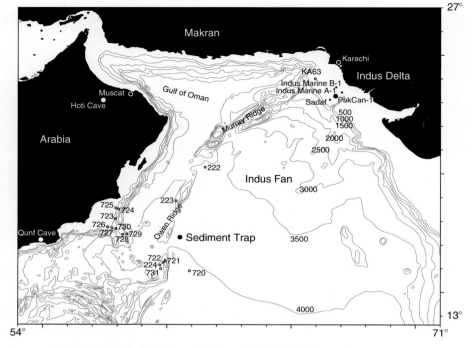

Figure 3.3 Bathymetric map of the Arabian Sea showing the major oceanographic features described in the text and the locations of Ocean Drilling Program and Deep Sea Drilling Project well sites. Water depth in meters taken from GEBCO compilation.

is a tendency to focus on the 8 Ma intensification it is noteworthy that the monsoon has not stayed at a constant strength since that time. The same study also showed periods of reduced upwelling at 5–6 Ma and, to a lesser extent, around 1 Ma.

Prell *et al.* (1992) were able to take this approach further and as well as looking at *G. bulloides*, also called attention to qualitative changes in the abundance of some radiolarians (a form of siliceous plankton) at roughly the same time (Figure 3.5). Furthermore, although bulk carbonate mass accumulation rates do not vary in phase with the abundance of upwelling fauna, strong coherent variations were noted in the accumulation of biogenic opal, specifically showing a peak at 10 Ma, but then a steady decline after 8 Ma. Opal accumulation, driven by production of siliceous plankton, does appear to be closely linked to nutrient supply in the modern oceans. Furthermore, some upwelling related radiolarian forms (e.g., *Collosphaera sp.*) showed an earlier blooming starting shortly after 12 Ma. Thus, the record of enhanced biogenic production in offshore Arabia provided the first, albeit rather complicated, image of monsoon intensification over long periods of geologic time.

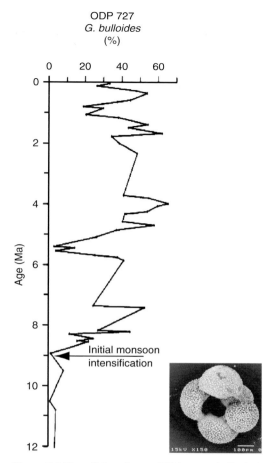

Figure 3.4 Plot of abundance of *Globigerina bulloides* in the Arabian Sea at ODP Site 727 on the Oman margin, showing the strong increase in upwelling strength after ~8 Ma (from Kroon *et al.*, 1991). Reproduced with permission of the Ocean Drilling Program.

3.2.2   Monsoon winds in the Eastern Indian Ocean

Since the early work on the Arabian Sea area, further definition of the evolution of the Indian Ocean monsoon has been achieved through analysis of cores from other areas. Gupta and Thomas (2003) provided evidence from ODP Site 758 on the Ninety-East Ridge that demonstrated increased strength of the north-east winter monsoon and greater seasonality starting around 2.8–2.5 Ma (Figure 3.6). Although not on the Arabian margin, this site is affected by upwelling driven by the summer south-west monsoon, which is the source of most of the organic flux and productivity in the southern Bay of Bengal. The 2.8–2.5 Ma age is close to the age of onset for Northern Hemispheric Glaciation

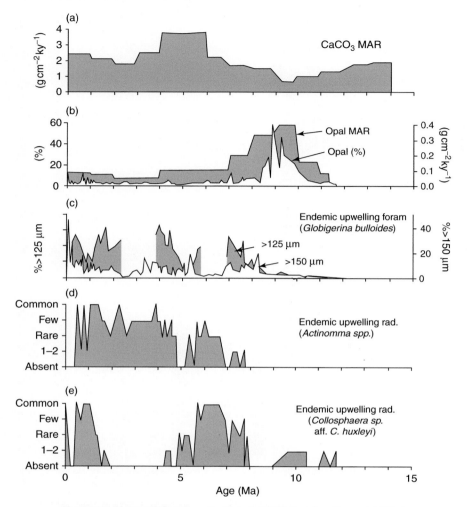

Figure 3.5 Plot modified from Prell *et al.* (1992) showing the variability in mass accumulation rates (MAR) in carbonate and opal, as well as the relative abundance of several upwelling related microfauna from ODP Site 722 on the Owen Ridge, showing the pattern of greater summer monsoon induced upwelling after 12 Ma. Reprinted with permission of the American Geophysical Union.

and demonstrates at a first-order level the links between monsoon intensity and the initiation of icehouse conditions elsewhere. These aspects are explored more fully in Chapter 4. Gupta and Thomas (2003) analyzed biofacies (i.e., assemblages of microfauna) to reconstruct paleoclimate during the Neogene. Before ~3 Ma three biofacies (Gs, Up and Sa, named after their primary species *Globocassidulina subglobosa, Uvigerina proboscidea, Stilostomella abyssorum*, respectively) alternated. These groups indicated low to moderate seasonality and year-round high

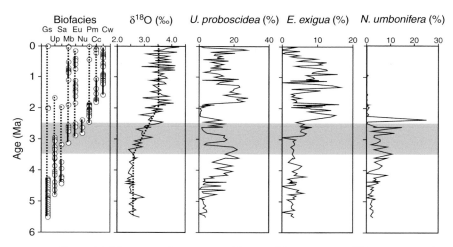

Figure 3.6 Monsoon paleoceanographic record from Ocean Drilling Program Site 758 on the Ninety-East Ridge from Gupta and Thomas (2003). Benthic foraminiferal biofacies (Gs – *Globocassidulina subglobosa*, Up – *Uvigerina proboscidea*, Sa – *Stilostomella abyssorum*, Mb – *Melonis barleeanum*, Ee – *Epistominella exigua*, Nu – *Nuttallides umbonifera*, Pm – *Pyrgo murrhina*, Cc – *Cassidulina carinata*, Cw – *Cibicides wuellerstorfi*), benthic foraminiferal oxygen isotope record (after Chen *et al.*, 1995, for 3.2–0 Ma), and percentages of key benthic foraminiferal species at Ocean Drilling Program Site 758. The gray bar indicates a time period between 3.2 and 2.5 Ma, when Northern Hemisphere ice sheets became established (Shackleton *et al.*, 1995; Maslin *et al.*, 2001). Reproduced with permission of the Geological Society of America.

primary productivity before 3 Ma. Seasonality is a key character of the monsoon and the data suggested a weak monsoon at 3–6 Ma. However, after ~3.0 Ma, six different assemblages alternated, which was interpreted to indicate greater seasonality and a low to medium flux of organic matter to the sea floor. Some species, such as *Epistominella exigua*, thrive on a strongly seasonal flux of organic material and, as a result, their blooming is taken to indicate more monsoonal-driven seasonality. In contrast, the observation of *Nuttallides umbonifera* indicates the presence of cold water, probably reflecting an increased flux of Antarctic Bottom Water at this time, as the ice sheets grew. This study showed that the increasing strength of the north-east monsoon and the decreasing strength of the south-west monsoon (i.e., more seasonality) at 3 Ma was coeval with enhanced Northern Hemisphere glaciation and affected the paleoceanography of the entire Indian Ocean. Gupta and Thomas (2003) hypothesized that fluctuations in the strength of the winter monsoon over the Bay of Bengal since 6 Ma might be linked to shifts in the latitudinal position of the westerly winds and the location of the Siberian high-pressure system (Chen and Huang, 1998), rather than being linked to surface uplift of the Tibetan Plateau, which is known to change on longer timescales than the climatic signals recorded. Unfortunately the Site 758 record did not extend far enough back in time to test whether 8 Ma was a key point of intensification in this region, as well as in the Arabian Sea.

3.2.3   *Monsoon winds in the South China Sea*

The onset date for monsoon intensification in the South China Sea has received much less attention than that in the Arabian Sea, partly because the South China Sea was drilled to great depths much more recently than its western equivalent, and partly because the monsoon has a much less profound effect on patterns of upwelling in the area. Figure 3.7 shows the bathymetry and

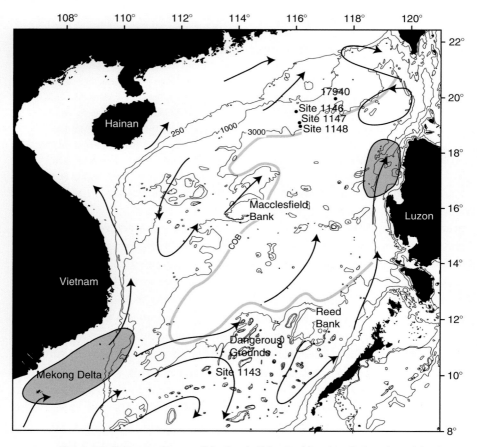

Figure 3.7 Bathymetric map of the South China Sea showing the location of the Ocean Drilling Program drill sites discussed and the areas of active summer upwelling, shown as gray-shaded regions. Lines with arrows show the direction of surface water currents during summer monsoon (Chu and Li, 2000). Bathymetry is in meters from GEBCO compilation. COB = Continent–Ocean Boundary.

regions of summer monsoon-induced upwelling in the South China Sea. These are concentrated offshore of the Mekong Delta and adjacent to Luzon island in the Philippines. In a study of core taken at Ocean Drilling Program Site 1143 in the Dangerous Grounds region of the southern South China Sea, Chen *et al.* (2003) examined the radiolarian fauna in order to reconstruct monsoon intensity, much as done by Prell *et al.* (1992).

Sediment trap data have demonstrated that increased flux of organic and siliceous matter to the seafloor in the northern South China Sea is mainly driven by enhanced productivity caused by upwelling during the winter monsoon. In contrast, in the central and southern South China Sea the maximum organic flux is closely related to the summer monsoon (Chen *et al.*, 1998). Variations in the abundance of radiolarians, especially Pyloniid forms, as well as radiolarian flux and species diversity are all considered good proxies of planktonic productivity and thus upwelling in the Indian and western Pacific Oceans (Heusser and Morley, 1997; Gupta, 1999). As in the Arabian Sea, it was assumed that those summer monsoon processes that cause strong productivity in the modern basin also drove production in the past.

The radiolarian paleo-monsoon proxies suggest that the East Asian summer monsoon was first initiated close to the Middle or Late Miocene boundary at ∼12–11 Ma and reached a maximum strength at ∼8.24 Ma (Figure 3.8). Decreased benthic foraminifer diversity after 3 Ma at ODP Site 1143 was interpreted by Hess and Kuhnt (2005) to reflect greater carbon flux to the region driven by a stronger winter monsoon after that time. As in the Arabian Sea, however, ODP Site 1143 only provided a limited record, dating back to ∼12 Ma. The period 8.8–7.7 Ma is marked out as being of especially strong summer monsoon-related upwelling. Interestingly the period after 7.7 Ma appears to be one of weak summer monsoon activity, supporting the record from the Arabian Sea that implied that this was a short period of strong summer monsoon. The apparent earlier onset to the East Asian monsoon compared with the Arabian monsoon further reinforces the idea that the two systems might be decoupled from one another on long timescales. As noted in Chapter 2, there are modeling reasons to believe that the south-west monsoon may be more critically dependent on the topography in western Tibet and the Himalayas than the East Asian monsoon (Chakraborty *et al.*, 2002), which is clearly also affected by the Western Pacific Warm Pool, since this affects the temperature differential between land and ocean in East Asia, but has no impact in the Indian Ocean.

3.2.4   *Regional and global influences*

Despite the apparent differences in the monsoon across South and East Asia the basic model of Kroon *et al.* (1991) and Prell *et al.* (1992), suggesting

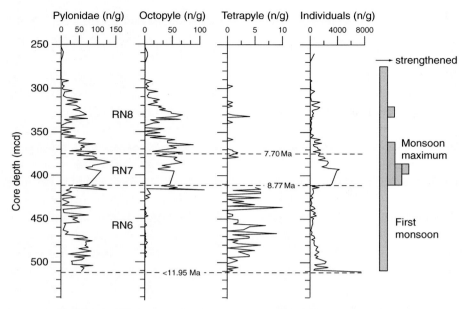

Figure 3.8 Abundance variations of *Pyloniid* radiolarians measured by Chen *et al.* (2003) at ODP Site 1143 in the southern South China Sea. These radiolarians prefer warm and saline environments with a cooler SST. The parallel changes in the number of *Pyloniid* and total radiolarian abundance suggest their upwelling origin driven by summer monsoons. Rapid increases in Octopyle (replacing Tetrapyle) over the RN6/RN7 contact and major changes in various proxies across the RN7/RN8 boundary suggest the development of an early monsoon at ∼12–9 Ma, an intensifed monsoon period between ∼9 and 8 Ma and a weakened monsoon period after ∼7.7 Ma. Reprinted with permission of Elsevier B.V.

that the Asian monsoon initially strengthened at ∼8 Ma, is still widely accepted. Nonetheless, consensus on the nature of climate evolution has yet to be achieved, especially what aspects of the monsoon can be assigned to regional factors, such as Tibet, and what is a local response to more global forcing functions, such as Northern Hemispheric Glaciation. Peterson *et al.* (1992) were one of the first to cast doubt on the 8 Ma event being a local intensification event linked to Tibetan uplift. Instead these authors noted that biogenic productivity increased over much of the Indian and Pacific Ocean at 8 Ma, implying that the signal noted on the Oman margin might not be a simple monsoon signal, but instead a local response to a global change.

Similarly, Gupta *et al.* (2004) used paleoceanographic productivity indices from benthic foraminifers sampled in scientific drill sites in the eastern and western equatorial Indian Ocean to show that even away from those areas affected by summer monsoon upwelling the ocean experienced enhanced productivity

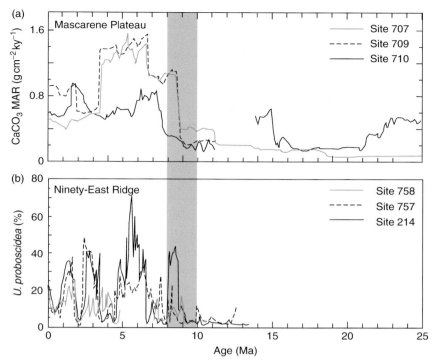

Figure 3.9 (a) Carbonate mass accumulation rates (MAR) in western equatorial Indian Ocean sites, Mascarene Plateau (ODP Leg 115), redrawn from Gupta *et al.* (2004). (b) Relative abundances of high productivity benthic foraminifer *Uvigerina proboscidea* plotted against age at ODP Ninety-East Ridge sites. Reproduced with permission of the Geological Society of America.

starting close to 8 Ma (Figure 3.9). Indeed this pattern of oceanic change at 8 Ma has now been recognized in the equatorial Pacific and even in the southern Atlantic, casting strong doubts as to whether this oceanographic event was really monsoon related at all, but could instead be only the local response to a global climatic change. Gupta *et al.* (2004) suggested that increased glaciation on Antarctica at that time may have strengthened temperature differences between high and low latitudes and thus global wind regimes, causing widespread open-ocean, as well as coastal, upwelling over a large part of the Atlantic Ocean and Indian and Pacific Ocean regions during the Late Miocene. This increased upwelling could have triggered the widespread increased biological productivity seen by Peterson *et al.* (1992) and Gupta *et al.* (2004). It is also noteworthy that there are few discernable tectonic events in the Himalayas or Tibet at this time that can be tied to monsoon strengthening (Chapter 2). It is also possible that other changes in the ocean–atmosphere system may also have played a role.

Around 10 Ma, the carbon isotope records of the Pacific and Atlantic started to diverge (Zachos *et al.*, 2001), and Roth *et al.* (2000) have suggested that the North Atlantic Deep Water first started to form at that time, resulting in a major re-organization of the global ocean circulation and upwelling system. Such a shift may have had wide-ranging climatic effects, extending to include the Asian region.

3.3   Continental climate records

3.3.1   *Weathering histories in the Western Himalayas*

Intensifying monsoon strength dominates the climate of continental South and East Asia and consequently controls the nature of continental weathering and erosion in this region, as well as the types of fauna and flora that can thrive there. Although continental sedimentary records are often poorly preserved and hard to date they do form a useful repository of information on the evolution of the Asian monsoon. Quade *et al.* (1989) used evidence from paleobotanical remains to argue for a major change in the flora and thus the climate of Pakistan during the Late Miocene. These authors used carbon isotope changes in carbonate concretions from soils to document a major shift from tree-dominated C3 to grass-dominated C4 types of vegetation in the Indo-Pakistan foreland regions around 7.5 Ma (Figure 3.10). As a result they suggested that this change was driven by monsoon intensification, because modern monsoonal weather patterns presently favor the growth of C4 grasslands. While the basic observation of a change from C3 to C4 domination appears to be robust, it was not clear whether this change was only monsoon dominated or whether it reflected a separate long-term global evolution in grass species. Indeed similar changes have been identified in North America, well outside the influence of the Asian monsoon. This observation led Cerling *et al.* (1993) to propose that the change from C3 to C4 grass types was rather a response to decreasing atmospheric CO_2 and global cooling at this time, not monsoon intensification.

As well as documenting the shift in vegetation type, Quade *et al.* (1989) measured the oxygen isotope character of carbonate concretions in soils from the same sedimentary sequence, which are interpreted to represent the composition of the ground waters from which they precipitated. In turn, those reflect the rainfall, whose oxygen isotope character is determined by the temperature at which precipitation occurred. The isotopic shift determined by Quade *et al.* (1989) seems to imply a strong warming of rain waters over Pakistan around 6–8 Ma. However, Prell *et al.* (1992) have hypothesized that instead of indicating a marked climatic warming this isotopic shift indicates a change of the dominant rain season from winter to summer time during intensification of the summer monsoon.

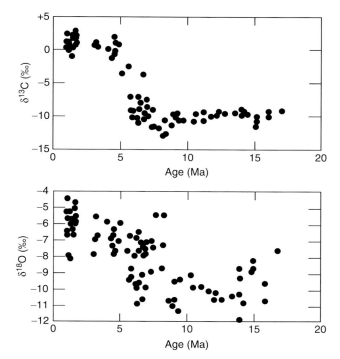

Figure 3.10 Diagram from Quade *et al.* (1989) showing the variations in $\delta^{13}C$ and $\delta^{18}O$ for pedogenic carbonates from the Siwalik Formation of northern Pakistan. Changes in $\delta^{13}C$ and $\delta^{18}O$ around 6–8 Ma are interpreted to reflect major change in rainfall and floral composition that may be linked to changes in monsoon intensity at that time. Reproduced with permission of Macmillan Magazines Ltd.

Attempts to repeat this isotopic reconstruction in other parts of the Himalayas have not been successful. Harrison *et al.* (1993) showed a very slight shift in oxygen isotopes around 8 Ma in Nepal, coinciding with the change in carbon isotope character, although it is not clear whether this precludes a warming at this time because the oxygen isotope character of rain in the Indus basin differs significantly from other parts of South Asia, including Nepal. These differences result in a potentially much more dramatic isotopic response to climate change in Pakistan than in other parts of South Asia.

Application of oxygen isotope methods to fossil-shell material in the western Himalayas can be used to assess the seasonality of the regional climate, with greater seasonality being interpreted as reflecting a stronger monsoon cycle. Dettman *et al.* (2001) used advance micro-sampling methods to extract tiny volumes of carbonate from mollusk shells found preserved in the Miocene Siwalik Formation sedimentary rocks now outcropping in northern Pakistan.

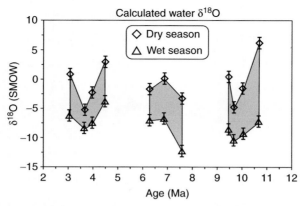

Figure 3.11 Diagram from Dettman *et al.* (2001) showing the changing seasonality of rainfall in Pakistan. The plot shows the estimated $\delta^{18}O$ for surface waters during dry and wet seasons derived from analysis of mollusk shells. Using maximum and minimum $\delta^{18}O$ values for shell carbonate and temperatures based on paleobotanical data, water $\delta^{18}O$ is calculated using fractionation relationship for molluscan aragonite. Estimated temperatures are different for three shaded areas: the oldest employs 20–3 °C; the middle uses a range of 17–32 °C; and the youngest interval uses 14–34 °C. Error bars represent a ±5 °C uncertainty in temperature used. SMOW is standard mean ocean water. Reproduced with permission of the Geological Society of America.

In particular, these workers were able to take carbonate from separate growth zones produced during both the summer and the winter seasons. Because the $\delta^{18}O$ value of the carbonate reflects both the $\delta^{18}O$ of the rainwater and the temperature at which precipitation occurred, the shell records can be used to assess the degree of seasonality between summer and winter. During the winter season, cooler temperatures, drier conditions and more positive $\delta^{18}O$ values for river water and rainfall must have occurred. In contrast, the most negative $\delta^{18}O$ values were assumed to reflect wet season conditions.

The results of these analyses are shown in Figure 3.11, demonstrating strong seasonality, i.e., a monsoonal climate back to at least 10.7 Ma. Unfortunately, a lack of suitable preserved material prevents extension of the record further back in time. Wet season $\delta^{18}O$ values of −9‰ to −10‰ imply rainfall of greater intensity than the current wet season prior to 7.5 Ma. The 2–5‰ increase in $\delta^{18}O$ values of bivalves, soil carbonates and mammal fossils starting around 7.5 Ma may instead reflect a drying of the climate around this time. This result is somewhat surprising because simple strengthening of the monsoon after 8 Ma, as suggested from the marine upwelling records, would have predicted stronger summer rains, not weaker, at that time.

3.3.2   Weathering histories in the Eastern Himalayas

Further evidence for continental climatic evolution in the Late Miocene comes from analysis of the clay minerals extracted from deep-water Ocean Drilling Program cores taken at Sites 717–719 in the SE Indian Ocean (Figure 2.7). Because these come from the Bengal Fan these materials are believed to represent the weathering regime in the Ganges-Brahmaputra basin at the time of their deposition, and can thus be used as climate proxies for that drainage. X-ray analysis of the clay reveals a change from an illite to a smectite-dominated assemblage around 7 Ma (Brass and Raman, 1991). Derry and France-Lanord (1996) used a battery of isotope methods to analyze the clay mineral fraction from these sediments in order to determine their origin and conditions of formation. Oxygen and hydrogen isotopes from the clay were compared with those known from the modern source regions and it was demonstrated that these had formed from weathering in the Ganges-Brahmaputra drainage, and crucially were not significantly altered as a result of their long transport offshore into the Indian Ocean.

Derry and France-Lanord (1996) were able to demonstrate a number of coherent changes in the clay record at ODP Sites 717–719 (Figure 3.12). It is clear that after 8 Ma there is a strong rise in the proportion of smectite relative to illite in the sediment, with a return to greater illite dominance again since 1 Ma. Smectite is typically interpreted to be the product of strong chemical weathering, while illite (and chlorite) are believed to reflect mechanical weathering in a metamorphic source terrain (Thiry, 2000). Smectite / (illite + chlorite) ratios are known to vary with monsoon strength on shorter timescales in the more recent past in the Mekong delta region (Liu *et al.*, 2004c). The clay minerals thus argue for more chemical weathering at 7.5–1.0 Ma, but does this imply more or less summer monsoon rain?

A clue to the answer of this question can be gained by consideration of the Sr isotopic character. $^{87}Sr/^{86}Sr$ values are not completely straightforward to interpret because they reflect both the provenance of the source regions and the intensity of chemical weathering in the basin. The carbonate rocks of the Tethyan Himalaya, located north of the High Himalaya, are estimated to provide much of the Sr in the Ganges-Brahmaputra ($\sim$70%), with $^{87}Sr/^{86}Sr$ values around 0.71, compared to values of 0.74 for the products of erosion and weathering of the crystalline High Himalaya. Thus the shift in $^{87}Sr/^{86}Sr$ values after 7.5 Ma may partially indicate relatively more weathering of crystalline sources at that time. However mass balancing arguments also indicate that the total flux of Sr from the Ganges-Brahmaputra must have decreased at the same time. Because of kinetic limitations on the rates of silicate weathering the shift documented in

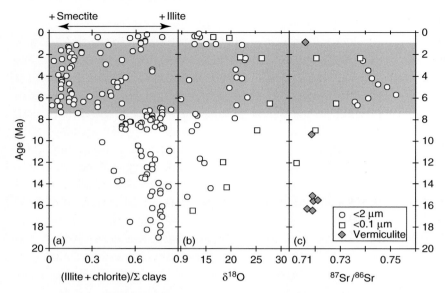

Figure 3.12 Reconstruction of evolving weathering regimes in the Ganges-Brahmaputra river basin from Derry and France-Lanord (1996). (a) Clay mineralogy showing the ratio of (illite + chlorite)/Σ clay as a proxy for the strength of physical erosion. (b) Oxygen isotope character of the clay minerals. The shift to lower δ^{18}O with increasing smectite content reflects more low-temperature weathering reactions. (c) ^{87}Sr/^{86}Sr of secondary minerals within the Bengal Fan, believed by Derry and France-Lanord (1996) to be a proxy record of ^{87}Sr/^{86}Sr of dissolved Sr in the river system and thus of the weathering regime in the flood plain. Reprinted with permission of Elsevier B.V.

^{87}Sr/^{86}Sr values requires an increase in the weathering environment at 7.5–1.0 Ma. The ^{87}Sr/^{86}Sr values of the clays formed as soil horizons reflect the ^{87}Sr/^{86}Sr values of the ground waters and the main Ganges-Brahmaputra River itself. Shifts to higher ^{87}Sr/^{86}Sr values may be driven by chemical breakdown of muscovite and biotite micas and, to a lesser extent, K-feldspars.

Taken together, these data were interpreted to reflect less run-off due to a drying environment in the fluvial basin at that time. Less run-off allowed eroded sediment to be stored for longer periods in the alluvial plain before being washed into the Indian Ocean, thus increasing the opportunity for chemical weathering provided the climate did not get too dry, so as to inhibit the chemical reactions of chemical weathering. As described in Chapter 5, sediments on the Bengal Fan were finer grained and accumulated at lower rates after 7.5 Ma, consistent with slower Himalayan erosion under a weaker summer monsoon environment. The recognition of similar ^{87}Sr/^{86}Sr patterns in foreland basin sedimentary rocks in Nepal by Quade (1993) suggests that these arguments are basically correct and that the deep-sea submarine fan can be used as a readily dated repository of information on weathering conditions in the continental drainage basin.

3.3.3   Weathering histories in Southern China

The weathering history of East Asia is known over the greatest time-scales from ODP Site 1148, which is located on the northern margin of the South China Sea (Figure 3.7). The drill site is located slightly east of the Pearl River (Zhujiang) estuary and has collected a hemipelagic rain, likely mostly from the Pearl River, together with limited input from Taiwan, Luzon and even wind-blown dust from the Loess plateau (Clift *et al.*, 2002a; Li *et al.*, 2003; Tamburini *et al.*, 2003). As a result, the site is well placed to record the changing erosional regime in southern China since opening of the basin during the Oligocene. Clift *et al.* (2002a) used standard X-ray based semi-quantitative analytical methods (Biscaye, 1965) to demonstrate a first-order subdivision of the erosion history. Prior to 15 Ma the proportion of smectite in the sediment was relatively high, albeit rather variable. However, since 15 Ma the clay mineral assemblages are dominated by illite and chlorite (Figure 3.13). This pattern was interpreted to reflect a general change from a variable climate with at least significant periods dominated by chemical weathering in the Oligocene–Early Miocene, while since 15 Ma weathering has been dominated by physical erosion.

Several other observations help put this pattern in context. The switch to a more erosive physical environment broadly correlates with faster clastic mass accumulation rates, suggestive of increasing run-off during the Middle Miocene (Clift *et al.*, 2004, Clift, 2006). Furthermore, Nd isotope analysis of the clay fraction indicates relatively limited evolution in the provenance since $\sim$25 Ma (Clift *et al.*, 2002a; Li *et al.*, 2003). It is noteworthy that although the upper reaches of the Pearl River now incise the edge of the Tibetan Plateau this river is almost unique as a major drainage in East Asia in not being sourced deep inside the plateau. The Pearl is also not considered to have been involved in large-scale drainage capture (Clark *et al.*, 2004). The relatively steady state character of the Pearl River means that changes in its erosional products offshore are likely linked to climate change rather than tectonic activity in the source terrains, or major headwater drainage capture.

Most recently, a more detailed analysis of the evolving climate at ODP Site 1148 was attempted by Clift (2006). In this study a higher resolution weathering record was generated using scanned Diffuse Reflectance Spectroscopy (DRS) data from the core, exploiting a technique developed by Giosan *et al.* (2002) who demonstrated that the 565 and 435 nm wavelength bands are sensitive to the presence of hematite and goethite respectively. A ratio of the 565/435 color

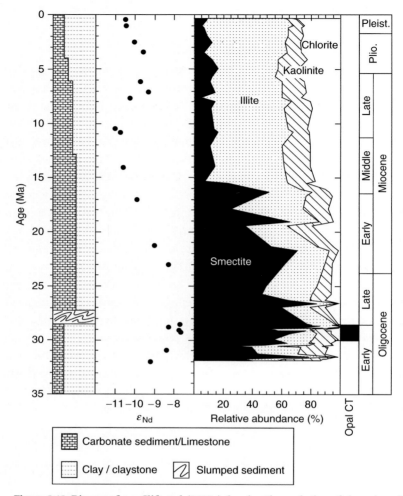

Figure 3.13 Diagram from Clift *et al.* (2002a) showing the evolution of clay mineralogy and clay Nd isotope character from ODP Site 1148 in the South China Sea. Location is shown on Figure 3.7. Note end of significant smectite influx to the basin prior to 15 Ma. Reprinted with permission of Elsevier B.V.

intensity can thus be used as a proxy for the relative abundance of these two minerals. Because hematite is preferentially formed in more humid conditions, while goethite is favored by more arid areas (Thiry, 2000) the color data can provide a first-order measure of weathering conditions in the drainage basin of the Pearl River. More detailed spectral analysis allows the relative proportion of chlorite to be isolated. Figure 3.14 shows that at 25–15 Ma there were a series

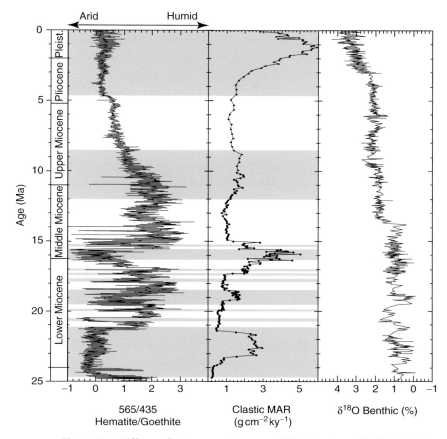

Figure 3.14 Diffuse reflectance spectroscopy (DRS) data from ODP Site 1148 from the northern South China Sea showing that the erosional flux from southern China, largely the Pearl River basin, has strong temporal variations in aridity since the start of the Neogene. MAR = mass accumulation rate. Oxygen isotope data is for benthic foraminifer *C. wuellerstorfi* (Wang *et al.*, 2003b). Modified after Clift (2006).

of variations between drier and wetter climates that sometimes corresponded to faster clastic mass accumulations rates. What is clear, however, is the shift to more chemical weathering and presumably a stronger summer monsoon starting around 22 Ma.

The resolution in identifying changes in clastic accumulations rates is limited by the age picks in the core and the correspondence may in fact be closer than this appears. Nonetheless, even with the presently available age control the periods 23.3–21.4 Ma, 19.3–18.5 Ma, 17.3–14.0 Ma and 12–9 Ma appear to be times of rapid clastic flux, falling within the period of rapid erosion reconstructed from the seismic profiles. The same pattern is seen in the Plio–Pleistocene when erosion was again rapid. The Pearl River drainage appears to become much wetter and more erosive after 22 Ma and 14–15 Ma, following the onset of Antarctic glaciation, shown clearly by the step in the benthic oxygen isotope record. This record suggests a more complex weathering history than previously reconstructed and is especially interesting in suggesting that the Early to early Late Miocene might have been a time of significant humid conditions. This period is broadly correlated with the global Mid-Miocene Climate Optimum (Zachos *et al.*, 2001). This record is important because it is also somewhat longer than those presently available in the Arabian Sea and the Bay of Bengal and provides potentially important new evidence that the East Asia summer monsoon may have strengthened much earlier than previously suggested, i.e., at least by 22 Ma.

3.4   Eolian dust records

3.4.1   *Eolian records in the Pacific Ocean*

An alternative but effective method for reconstructing monsoon wind strength and continental aridity has been to examine the character of wind-blown dust carried by the monsoon (Nair *et al.*, 1989). This dust accumulates in oceanic settings where the material cannot have been deposited by the normal mass wasting processes that operate on continental margins. As well as the dust blowing from Arabia into the northern Indian Ocean (Figure 3.1), the deserts of central Asia are an important source of wind-blown dust, which is typically transported toward the east during the months of the East Asian winter monsoon (Figure 3.15). This dust is partially deposited in northern China in the Loess Plateau, but large volumes are also advected far into the Pacific Ocean, even reaching the Americas. In an attempt to quantify the reach of this dust plume, Rea (1994) used samples from the youngest core tops at a series of locations across the Pacific Ocean to map the distribution of wind-blown material. This type of analysis required the physical and chemical separation of the eolian fraction from the biogenic mass of the sediment. In most locations far into the Pacific Ocean, especially those located on the tops of ridges or seamounts, the wind is the only viable agent to bring such material. Figure 3.16 shows a map of the eolian accumulation pattern for the Holocene in the Pacific Ocean. What is clear is that the highest rates ($>$1000 mg cm^{-2} ky^{-1}) are located directly east of China and thin both east and south from there, reflecting the source and the direction of the prevailing winds. In comparison, the South Pacific is relatively free of eolian sediment. As a result eolian dust records in the North Pacific can be used to track the intensity of East Asian monsoon winds.

The source of the eolian dust in the Pacific Ocean has been investigated using isotopic fingerprinting methods. Jones *et al.* (1994) showed that modern North

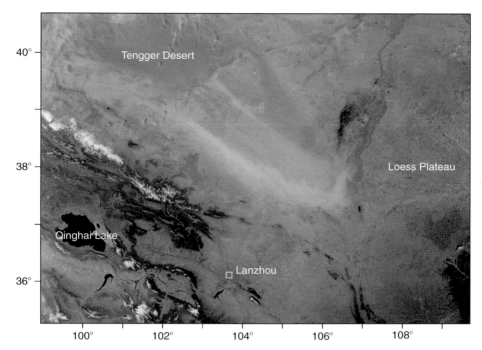

Figure 3.15 NASA satellite image (MODIS from Visible Earth) of the Tengger Desert in north-west China showing a powerful winter dust storm putting large volumes of material into the atmosphere, which is being transported to the south-east. Note: Qinghai Lake and the upper reaches of the Yellow River are in the bottom left hand corner. Images reproduced with permission of NASA. See color plate section.

Pacific dust was identical in Nd and Sr isotopes to the Chinese Loess. Pettke *et al.* (2000), studied dust material spanning the last 12 Ma extracted from cores recovered from ODP Sites 885 and 886, located just south of the central Aleutian Arc. These were compared with dust samples from Central Asia and lavas from Kamchatka, which are a possible major source of tephra, and also located upwind of the drill sites. Using a combination of Nd isotope ratios (recalculated as the factor ε_{Nd}) and measured $^{238}U/^{204}Pb$ ratios (μ_2), Pettke *et al.* (2000) argued that the vast majority of the eolian dust in the North Pacific is of Asian origin and can thus be used to examine the varying strength of the winter monsoon (Figure 3.17). Moreover, the lack of any coherent change in Pb or Nd isotopes since 12 Ma indicated that the sources had not changed significantly since that time, while the isotope ratios themselves provided a good match between the Pacific dust and the continental loess exposed in China.

Temporal evolution in the chemistry and mineralogy of ODP Sites 885 and 886 has also been documented in terms of $^{87}Sr/^{86}Sr$, as well as the ratios of smectite

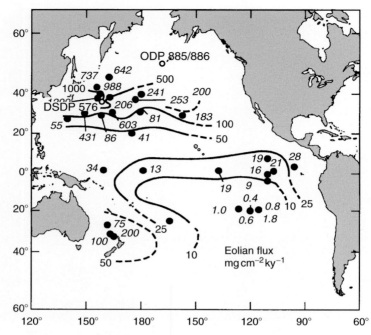

Figure 3.16 Map from Rea (1994) showing the variations in mass accumulation rate of dust in the surface sediments of the Pacific Ocean. The values usually represent the uppermost Quaternary sample available and occasionally are an average of Holocene rates. East Asia is the largest source of dust to the Pacific. Note locations of DSDP Site 576 and ODP Sites 885/886. Reprinted with permission of the American Geophysical Union.

to plagioclase or illite (Figure 3.18). These proxies can be used to trace the intensity of chemical weathering in the source regions, similar in approach to that described from the Indian Ocean submarine fan clays. It is clear that the relative proportion of smectite increases after ∼7–8 Ma, indicating stronger chemical weathering after that time. The degree of chemical weathering would seem to fall again after ∼3 Ma, although it is important to note that, unlike the sediment older than 8 Ma, there is now no kaolinite present (Pettke *et al.*, 2000). This pattern of mineral evolution suggests a progressive aridification of Central Asia, since 12 Ma, but with significant steps in the process at 3 Ma and 8 Ma. Prior to 8 Ma, Central Asia was wet enough to have significant physical erosion, while the aridity has intensified to such an extent since 3 Ma that little chemical reaction can occur in the area, which is dominated by physical erosion in flash flood events and eolian transport.

Further information about monsoonal paleoclimate and monsoon wind intensity can be gained from calculating the mass accumulation rates of eolian

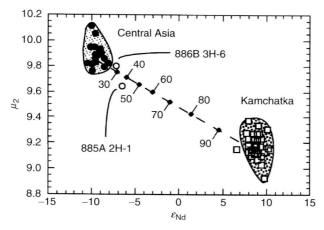

Figure 3.17 ε_{Nd} versus μ_2 values (measured $^{238}U/^{204}Pb$ ratio) plotted for Kamchatka lavas (open squares) and the central Asian dust (full circles) from Pettke *et al.* (2000) demonstrate the huge difference in their ε_{Nd} values and their significantly different unradiogenic Pb signatures. Full diamonds along the tie-line show the calculated wt% of a Kamchatka-type admixture in a bulk dust sample. Two dust samples (open circles) may be contaminated by up to 35 wt% Kamchatka-type material, which is in line with core log reports from these sites (Rea *et al.*, 1993). Note that contributions of up to 15 wt% volcanic arc component cannot be resolved in Nd–Pb isotopic space owing to minor source heterogeneities in central Asia. Reprinted with permission of Elsevier B.V.

material and determining the grain size of the material. Typically the mass accumulation rate is thought of as representing the supply of fine material, and can thus be used as an aridity index because drier regions tend to yield more eolian dust than humid ones (e.g., Prospero *et al.*, 1989). In contrast, the grain size variation at any given location can be used to assess changing wind energy and thus speeds, with coarser grains carried by stronger sustained winds (e.g., Janecek and Rea, 1985). Although the basic premise is simple, some caution needs to be exercised when considering long periods of time at a drill site that may be moving relative to the eolian source regions as a result of plate motions. In the case of the Pacific Plate, motions since 45 Ma have been toward the WNW, bringing any given drill site closer to the sediment sources and potentially increasing grain size with no actual change in wind speed. Fortunately rapid climatically driven changes are relatively easy to distinguish from this long-term background trend.

An example of the application of this approach is shown from the north-west Pacific Ocean, where the variation in grain size and mass accumulation rates since 80 Ma have been reconstructed (Figure 3.19). At Deep Sea Drilling Project (DSDP) Site 576 the eolian mass accumulation rate is relatively flat and low from 80 Ma to 10 Ma, but shows rapid increase after ~4 Ma, indicative of sharply

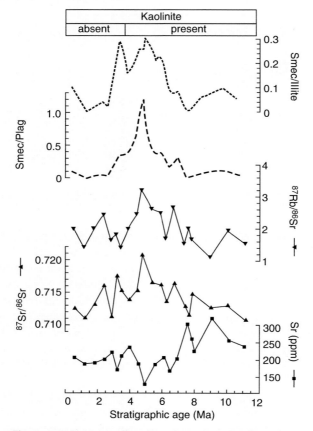

Figure 3.18 Data on eolian dust plotted as a function of depositional age from
Pettke *et al.* (2000). From bottom to top: measured Sr concentrations ([Sr] in ppm; filled
squares), Sr isotopic compositions (^{87}Sr/^{86}Sr; filled triangles) and ^{87}Rb/^{86}Sr ratios
(filled inverted triangles). Dust mineralogy for the 2–20 μm grain size fraction is
represented as three-point smoothed data; smectite vs. plagioclase (Smec/Plag;
long-dashed) and smectite vs. illite (Smec/Illite; short-dashed), correlated with the
presence or absence of kaolinite. Reprinted with permission of Elsevier B.V.

increased continental aridity after that time. In contrast, the grain size shows
substantial variability, with a sharp fall around 55 Ma indicating much slower
winds after the major climatic re-organizations around the Paleocene–Eocene
boundary (∼55 Ma). Of greater importance to our understanding of the Asian
monsoon evolution is the increase in grain size after ∼30 Ma, peaking around
25 Ma, then falling slightly before again strengthening after 10 Ma. This pattern
of development suggests that the winter monsoon winds started to blow more
strongly after ∼30 Ma, weakened at around 10 Ma before peaking again in the
Late Pliocene–Early Pleistocene.

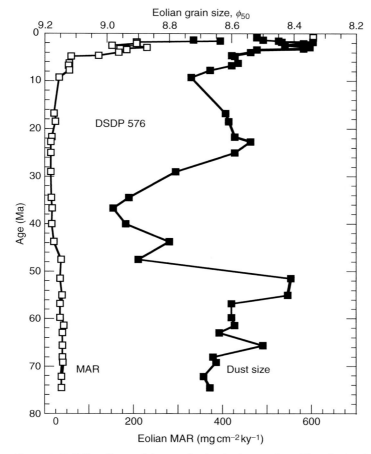

Eolian grain size, ϕ_{50}

Age (Ma)

DSDP 576

MAR

Dust size

Eolian MAR (mg cm^{-2}ky^{-1})

Figure 3.19 Eolian flux and dust grain size in the North Pacific calculated by Rea (1994) for DSDP Site 576. Note the large change in flux about 3 Ma and the large change in grain size at around 55 Ma. Reprinted with permission of the American Geophysical Union.

A more detailed image of Middle Miocene–Recent monsoon development was provided by a study by Rea *et al.* (1998) of a core from ODP Site 885/886 in the North Pacific, which receives most of its eolian flux from Central Asia (Figure 3.20). At this location the eolian mass accumulation rate shows distinct spikes to higher values at 7.5–8.0 Ma and 5.5–6.5 Ma, as well as after 3.5 Ma. These periods are taken to represent periods of higher continental aridity, which since 3.5 Ma is clearly correlated with the intensification of Northern Hemispheric Glaciation and general global climatic deterioration (Zachos *et al.*, 2001). It is noteworthy that, as at DSDP Site 576, there is a distinct decoupling between grain size and eolian mass accumulation rate, with a steady increase in grain

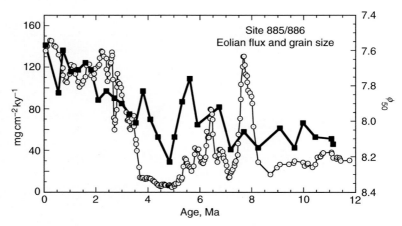

Figure 3.20 Eolian flux and dust grain size in the Northwest Pacific calculated by Rea (1994) for ODP Sites 885/886. Note peaks in eolian mass accumulation rates after 3.5 Ma correlating with onset of Northern Hemispheric Glaciation. Reprinted with permission of the American Geophysical Union.

size being recorded since 4.5 Ma, indicating strengthening winter monsoon winds since that time. Initial strengthening of the winter monsoon appears to occur after 8 Ma, weakening again after around 5.5 Ma. This pattern mirrors that found in the *G. bulloides* upwelling record of the Arabian Sea (Figure 3.4; Kroon *et al.*, 1991). What is clear is that there is little evidence for a simple one-stage monsoon intensification starting around 8 Ma in the Pacific dust record.

3.4.2   Eolian records in the Chinese Loess Plateau

As well as blowing dust from the deserts of Central Asia far out into the Pacific Ocean, winter monsoon winds also carry and deposit significant volumes onshore in East Asia, in Siberia, in Japan, but especially in the Loess Plateau of northern China (Figure 3.21). Although loess is found throughout China, the thickest and more complete sections have accumulated in central northern China, south and east of the Tengger Desert and the Ordos Basin, just north of the Qilian Shan, and in a region now incised by the Yellow River (Figure 3.22). The Loess Plateau differs significantly from the surrounding deserts in being a region of deposition, even if it is not substantially wetter (rainfall ~400 mm y^{-1}). The soft, poorly consolidated character of the sediments makes them susceptible to rapid erosion, which has been exacerbated in historical times by the intensive agricultural development of the region. Indeed, it is the erosion of loess into the river that gives the Yellow River its name. The geomorphology of the region is one of flat plateau regions, deeply incised by steep gullies (Figure 3.23). The geology of the loess is characterized as alternating series of light-colored loess

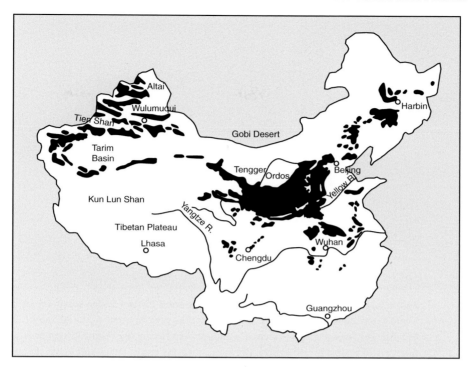

Figure 3.21 Schematic map of China showing regions of thick loess accumulation (>10 m) in black. The Loess Plateau region is the major black region in central northern China, south of the Tengger Desert and Ordos Basin (from Wang, 2004). Reprinted with permission of the American Geophysical Union.

sediments and darker red-colored paleosols dating back into the Late Pliocene (Ding *et al.*, 1999, 2001; An *et al.*, 2001; Guo *et al.*, 2002). These layers are typically interpreted to reflect alternating periods of stronger and weaker winter monsoon. Paleomagnetic dating has shown that most loess sequences date back to ∼8 Ma, although with a significant increase in accumulation rates since ∼2.7 Ma (e.g., Ding *et al.*, 1999).

As in the marine sequences from the North Pacific described above, the records of monsoon intensity and continental aridity preserved in the loess sediments can be deciphered by looking at mass accumulation rates and grain size. In addition, magnetic susceptibility and geochemistry have been employed to understand the climatic signals in the sediments. Magnetic susceptibility is believed to depend mostly on ultra-fine-grained ferromagnetic minerals formed *in situ* during pedogenesis (Maher, 1986). Therefore, susceptibility records in the loess sequences can be used as indices of summer monsoon precipitation because it is wetter conditions that promote soil formation (Kukla *et al.*, 1990).

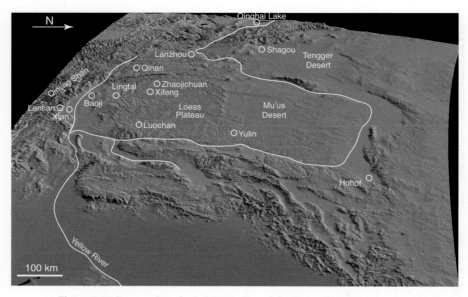

Figure 3.22 Perspective, shaded DEM view of the Loess Plateau region created from GeoMapApp showing the physical geography of the area adjacent to the NE flank of Tibet. The Loess Plateau is incised around its exterior by the Yellow River, but is otherwise a relatively arid region. The map shows the major towns and location of the best-studied loess sections mentioned in the text.

Drier conditions instead favor accumulation of loess. Furthermore, the overall strong correlation of magnetic susceptibility records with an independently derived Rb/Sr time series (Chen *et al.*, 1999) demonstrates that both indices are measures of summer monsoon strength.

Some caution does however need to be exercised with loess sequences because traditionally these have been interpreted as relatively unbroken climatic records. Now high-resolution optically stimulated luminescence (OSL) dates from China suggest that sedimentation is episodic at subglacial–interglacial time-scales. Sedimentation rates are seen to vary rapidly within units and between sites, including the generation of unconformities (Stevens *et al.*, 2006). All these factors affect the use and fidelity of the loess as a climate record.

During weathering, Rb is relatively stable, whereas Sr is quite mobile in aqueous fluids; therefore, an increased Rb/Sr ratio should indicate increased weathering and pedogenesis, and a strong summer monsoon. An example of the type of records derived is shown in Figure 3.24 showing the variability in grain size and magnetic susceptibility at Zhaojiachuan spanning ∼8 Ma (An *et al.*, 2001). The millennial-scale history of the loess will be discussed more in Chapter 4, but here we wish to note the sharp increase in the variability and average values of

Figure 3.23 (a) Aerial photograph of typical terrain in the Loess Plateau of China showing the deep gullies and ravines between plateau regions within which the loess–paleosol sequences are now exposed. The flat plateau regions are now exploited as farm land. (b) Outcrop photograph showing the color alternations associated with paleosol–loess interbeds. (c) Overview of the famous Lingtai section.

the >0.19 mm grain size fraction in sediments younger than ~2.7 Ma. This pattern suggests that the intensity of the strong winter monsoon winds strengthened greatly at that time and that since then there has been significant shorter-term variability, likely linked to orbital scale processes and the intensity of Northern Hemispheric Glaciation. In contrast, the magnetic susceptibility of the loess does not increase in its average values at the same time, suggesting little average change in summer precipitation. Nevertheless, it is noteworthy that the degree of short-term variability changes sharply after 2.7 Ma. This implies that the degree of weathering and pedogenesis, driven by summer monsoon rains, is strongly variable on millennial-scale cycles of alternating wet and dry phases.

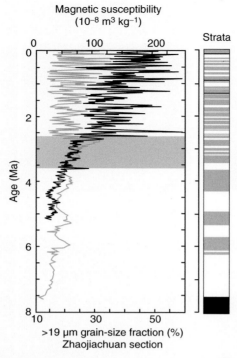

Magnetic susceptibility
(10⁻⁸ m³ kg⁻¹)

Figure 3.24 Terrestrial and marine records from the Chinese Loess Plateau modified from An *et al.* (2001). The shaded zone at 3.6–2.6 Ma indicates times of change. The stratigraphy and time series of magnetic susceptibility (thin line) and >0.19 mm grain-size fraction (thick line) from the Zhaojiachuan section on the Loess plateau. The chronology is based on the polarity boundary ages, and was obtained by interpolation from a sedimentation model using the >19 mm grain-size fraction and a magnetic susceptibility model in the basal part where grain-size data are not available. The white, gray and dark shaded patterns in the simplified stratigraphy column represent loess or loess-like sediment, paleosols and bedrock, respectively. Reproduced with permission of Macmillan Magazines Ltd.

More recently, An *et al.* (2005) have examined two long loess sections (Lingtai and Lantian) in order to reconstruct the evolving monsoon and flora of central Asia since the Late Miocene. As at Zhaojiachuan, magnetic susceptibility maxima and the development of paleosols are interpreted to indicate periods of stronger, wetter summer monsoon. At the same time this study analyzed the carbon isotope character of both carbonates in the soil horizons and from organic matter in the soils in order to constrain the nature of the flora that thrived during each episode of sedimentation. As in the Pakistan foreland basin, the relative carbon isotope character of the plant matter and the soil carbonate concretions is interpreted to reflect the balance of C3 to C4 plant types. The C4 plant abundance estimated from the $\delta^{13}C$ values of soil organic matter is consistent with the C4 biomass proportion obtained from living plants in the field.

An *et al.* (2005) compared the $\delta^{13}C$ data from the loess with various climatic variables and concluded that the estimated C4 biomass correlated positively with increased precipitation during the growing season and with the annual precipitation. The abundance of C4 plant biomass was also observed to be positively correlated with temperature, though less so than with precipitation. Thus a higher ratio of the C4/C3 plant biomass is favored by increased warm season rainfall in semi-arid regions with high seasonal contrast. This pattern exists because of the high water efficiency of C4 plants (Huang *et al.*, 2001). The section at Lingtai in the central western Loess Plateau (Figure 3.25) provides a climatic and faunal record extending back to 7 Ma. Soil carbonate $\delta^{13}C$ values of around $-6.7‰$ at 7 Ma in this location show that C4 plants accounted for ~38% of biomass at that time, which is approximately consistent with synchronous data from South Asia (Cerling *et al.*, 1997; Morgan *et al.*, 1994; Quade *et al.*, 1989). The first major shift in $\delta^{13}C$ values at Lingtai is noted at around 4.5 Ma and reached a peak at ~2.7 Ma as the proportion of C4 plants rose to an estimated 60%. Correlation of soil carbonate $\delta^{13}C$ values with magnetic susceptibility shows that the summer monsoon was at maximum strength over the Loess Plateau at 2.7 Ma, but then subsequently weakened to reach a base value at 2.0–1.5 Ma, as the climate became colder and drier.

It is noteworthy that the rate of eolian sedimentation on the Loess Plateau closely parallels that in the North Pacific described by Rea *et al.* (1998). This is to be expected because both locations are fed by dust derived from the progressive desiccation of central Asia. It is clear that summer monsoon strength proxies do not parallel or even mirror those for aridity and winter monsoon wind (Figure 3.25). The synchronous strengthening of the summer monsoon and greater continental aridity at ~3.5–2.7 Ma could reflect a stronger summer monsoon owing to a higher Tibetan Plateau, which in turn would increase the rain shadow effect in central Asia. The fact that these factors do not typically coincide in this fashion suggests that the relationships between summer and winter monsoon are not so simple.

The last 4 My of monsoon history are shown in greater detail from the loess section at Lantian on the southern edge of the Loess Plateau (Figure 3.26). Three intervals of stronger summer monsoon activity were identified on the basis of $\delta^{13}C$ analyses and magnetic susceptibility at 3.7–2.7 Ma, 1.5–0.9 Ma, and 0.6 Ma–present. It is, however, noteworthy that the periods of stronger summer monsoon are not continuously high but rather show strong variability linked to glacial–interglacial cycles. The higher magnetic susceptibility and stronger

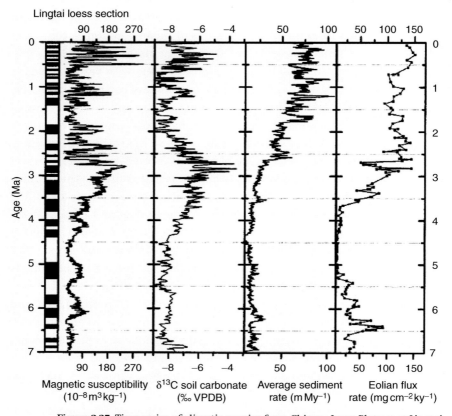

Figure 3.25 Time series of climatic proxies from Chinese Loess Plateau at Lingtai and comparison with eolian dust flux in North Pacific Ocean modified from An *et al.* (2005). Stratigraphic column of Lingtai section (288.5 m thick) for past 7.2 My. From left to right the columns show: magnetic susceptibility, soil carbonate $\delta^{13}C$, average dust sedimentation rate (SR) on Chinese Loess Plateau, based on six eolian sequences at Lingtai, Xifeng-Zhaojiachuan, Chaona, Luochuan, Jiaxian and Lantian, eolian flux from Ocean Drilling Program Sites 885 and 886 in North Pacific Ocean (Rea *et al.*, 1998). Timescales for Lingtai and Lantian sections were constructed using magnetic reversal boundaries (Sun *et al.*, 1997; 1998) as age controls. Reprinted with permission of Elsevier B.V.

summer monsoon are generally linked to interglacial periods. The Lantian record, like that at Lingtai, shows a strong correlation and coupling between the strength of the East Asian summer monsoon circulation and the composition of the biomass. An *et al.* (2005) concluded that in East and Central Asia, large-scale Late Miocene C4/C3 vegetation changes have been primarily driven by warm seasonal precipitation and temperature variations associated with changes in monsoon circulation.

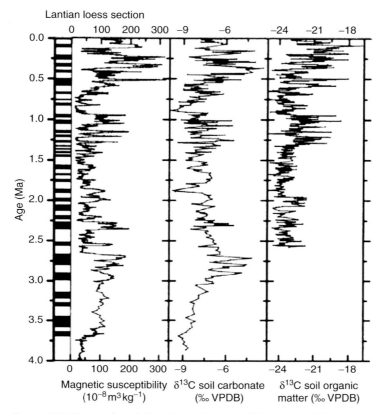

Figure 3.26 Time series of climatic proxies from Chinese Loess Plateau at Lantian modified from An *et al.* (2005). From left to right the columns show: stratigraphic column of Lantian section (163.7 m thick) since 4 Ma, characterized by alternating light-colored to reddish-yellow loess or silty loess and light red to brownish-red paleosol layers. Black bars indicate soil layers and empty bars indicate loess layers; magnetic susceptibility (SUS) of Lantian section. VPDB – Vienna Peedee belemnite; soil carbonate $\delta^{13}C$ from Lantian, sampled at average 40–50 cm interval; soil organic matter $\delta^{13}C$ from Lantian. Reprinted with permission of Elsevier B.V.

3.4.3   Onset of Loess sedimentation

As described above, the bulk of the loess–paleosol sequence has been deposited since 2.7 Ma, with much of the rest postdating 7 Ma. It might thus seem as though major continental aridification and strong winter monsoon winds date from this time too. However, there is evidence from the much longer North Pacific dust record that eolian grain sizes first began to increase around 25–30 Ma (Figure 3.19). A clue to earlier climate change and the start of loess sedimentation can be gained from the deposits underlying the loess–paleosol sequences, the so-called "Red Clay" or "Hipparion–Red Earth" deposits (so called because of their color and the abundance of Hipparion fossils; Liu, 1985). The Red Clay Formation ranges from 30 to 60 m thick over most of the Loess Plateau, but in some regions, such as at Lingtai, > 100 m are recorded (Ding *et al.*, 1999; Sun *et al.*, 1998). The nature of the Red Clay has been a source of controversy in the past. Andersson (1923) indicated that the Red Clay was the product of weathering of the underlying limestone, although more recently paleomagnetic and chemical arguments have been used to show that this is actually largely of eolian origin, and can thus be used to study the monsoon over longer timescales (Liu *et al.*, 1988; Ding *et al.*, 1998; Guo *et al.*, 2002). Similarities between the major, trace and rare earth element compositions of the loess layers within the Red Clay Formation and the younger loess suggest that their source regions are the same, or at least similar in composition, while the range of grain sizes further reinforces the hypothesis that the Red Clay Formation is also an eolian sediment, produced under similar wind strength conditions.

Paleomagnetic measurements together with mammalian fossils date the Loess–Red Clay section exposed around Qinan in the south-west of the Loess Plateau from around 22 to 6.2 Ma, with only minor hiatuses (Figure 3.27). The detailed reconstruction of a complete reversal record over that time interval reinforces the suggestion that the Red Clay Formation preserves a long Miocene record of winter and summer monsoon intensification. Guo *et al.* (2002) reported 231 soils and eolian layers visible in a sequence covering 15.1 My. This yields an average duration of each loess–soil pair of 65 000 years, which is of the same order as typical orbital-scale variations and suggests that Milankovich cyclicity may have been an important control on monsoon strength prior to the onset of icehouse conditions in the Pliocene. It should be noted that the long-term average dust accumulation rate is ~1.67 cm ky^{-1} in the Red Clay, much less than the >30 cm ky^{-1} noted since 2.7 Ma. Nonetheless, this is an important indication of monsoon wind activity in Central Asia dating back to at least 22 Ma.

3.5   Evolving flora of East Asia

The evolving flora of East Asia can be used as a proxy for monsoon development during the Cenozoic. The changing C3/C4 ratios of the biomass have been reconstructed from the carbon isotope character of the organic matter preserved in the soils of the Loess Plateau, as described above. A longer-term record of floral evolution can also be reconstructed using pollen data from the sedimentary basins. Pollen are distinctive of their source plants and are also very resilient to preservation in sedimentary rocks, thus making them valuable sources of paleoclimatic information.

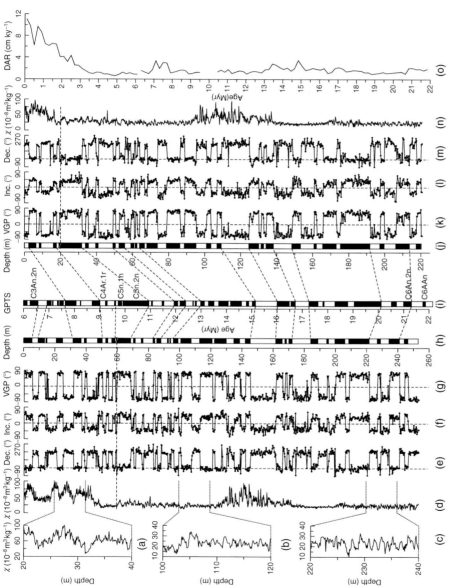

Figure 3.27 Magnetostratigraphy and magnetic susceptibility of the Qinan eolian sequences and Late Cenozoic dust accumulation rate in northern China as measured by Guo *et al.* (2002). (a)–(h), Profiles of magnetic susceptibility (χ), declination (Dec.), inclination (inc.), virtual geomagnetic pole latitudes (VGP) and polarity zonation of QA-I. (i) Reference geomagnetic polarity timescale (GPTS). (j)–(n) Polarity zonations, VGP, Inc., Dec. and susceptibility profiles of QA-II. (o) Dust accumulation rate (DAR) in northern China. Data for 0–6.2 Ma interval are calculated based on the loess–soil sequence and the Hipparion Red-Earth Formation at Xifeng. The 6.2–22.0 Ma portion is based on QA-I. Calculation was made at 200 000 year intervals. The timescales are obtained using geomagnetic boundaries as age controls and then interpolating, using the magnetic susceptibility age model for loess. Horizontal discontinuous lines indicate the erosion surfaces in both sections. Reproduced with permission of Macmillan Magazines Ltd.

83

A long, relatively continuous record has been reconstructed from the assemblages of pollen types in the Qaidam Basin of north central China, which lies just south of the Loess Plateau and the Qilian Shan (Figure 2.3). Because the basin lies adjacent to the northern edge of the Tibetan Plateau it can be a sensitive measure of surface uplift and resultant climate change. The sediments of the Qaidam Basin are continental in facies and are thus harder to date than their marine equivalents. Liu *et al.* (1990, 1988) and Yang *et al.* (1992) constructed a geomagnetic stratigraphy for the Tertiary and Quaternary strata in the basin and also used pollen and ostracod assemblages to determine an absolute age for each depositional package. Figure 3.28 shows how the different pollen assemblages

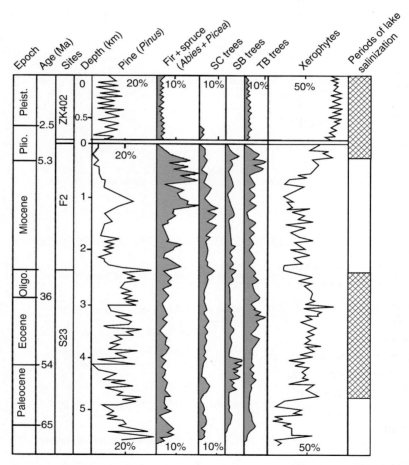

Figure 3.28 Cenozoic pollen sequence of Qaidam Basin and periods of salinization from Wang *et al.* (1999c). SC trees = subtropical conifer trees; SB trees = subtropical broad-leaved trees; TB trees = temperate broad-leaved trees. Reprinted with permission of Elsevier B.V.

change through time, as reconstructed by Wang, J. *et al.* (1999c). In particular, xerophytic sporo-pollen populations can be used to track the desiccation of the region because these types of pollen are produced by plants that thrive in arid conditions. Xerophytic sporo-pollen proportions are relatively sparse in the Early Paleocene (5–40%) and in the Miocene (20–50%) strata. However, relatively high percentages of xerophytic sporo-pollen were present from the Late Paleocene to the Late Oligocene (40–70%). The highest percentage of xerophytic sporo-pollen appeared in the Pliocene and Pleistocene (70–90%).

These data indicate that a more humid climate existed in the Early Paleocene and in the Miocene, while a relatively dry climate prevailed in the Qaidam Basin during the Late Paleocene and Late Oligocene. The Pliocene and Pleistocene stand out as the driest periods and also potentially the coldest, because subtropical broadleaf and conifer trees also disappear during the Pliocene. Other times of significant change are in the Early Miocene, when the proportion of pine pollen fell dramatically, and in the Middle Miocene, when the contribution from fir and spruce trees increased. Because these species are usually associated with cooler, higher elevations Wang *et al.* (1999c) interpreted their increase to reflect regional surface uplift around the Qaidam Basin. The subsequent fall in fir and spruce pollen during the Pliocene again reflects the drying of the region following the onset of Northern Hemispheric Glaciation and the uplift of the surrounding mountains and the Tibetan Plateau, causing a rain shadow to form.

Wang *et al.* (1999c) suggest that in the Qaidam Basin relatively dry climates prevailed in the Eocene and Oligocene and that these were caused by the location of the Subtropical High Pressure Zone. This situation may have changed as a result of a 600 km northward drift of western China during the Cenozoic, or because of monsoon intensification around the Oligocene–Miocene boundary (~24 Ma). The evolving flora of the Qaidam Basin can be understood in terms of changing monsoon climate driven by this drift and the uplift of the Tibetan Plateau.

A recent synthesis of pollen types and climatically sensitive sedimentary facies by Sun and Wang (2005) allows us to constrain further the evolving monsoon climate of the region covering modern China. In this work the occurrence of evaporites was interpreted to reflect dry and potentially warm climates, while development of oil shales and coals were linked to humid, warm conditions. The results of this analysis for the Linxia Basin in the western Tarim Basin are shown in Figure 3.29. In the Linxia pollen profile, herbaceous pollen declined significantly into the Miocene compared to their abundance in the Oligocene. At the same time, the pollen of needle-leaved conifers and broadleaved trees increased markedly between 22 and 8.5 Ma, indicating that forest vegetation replaced the drier forest or steppe, and that a warm and humid climate developed during the Early–Middle Miocene. Such a shift is consistent with a strengthening

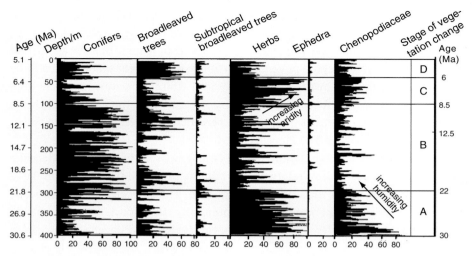

Figure 3.29 Late Cenozoic percentage pollen diagram from Linxia in the western Tarim Basin, modified from Ma *et al.* (1998).

summer monsoon at this time. Subsequently, Ma *et al.* (1998) reported a rise in the relative importance of herbaceous pollen and the synchronous decrease in tree pollen at 8.5–6.0 Ma, suggesting a developing steppe vegetation and reduced temperature and humidity. The flourishing of *Chenopodiaceae*, a pollen type associated with continental aridity is consistent with a weakening summer monsoon after 8.5–6.0 Ma in western China compared with the more humid environments of the Early and Middle Miocene. Most recently, since 6.0 Ma, the herbaceous pollen values fell and the tree pollen increased again, ending the cool and humid climatic conditions of the Miocene, and replacing them with the much more arid conditions seen in the region to this day. Similar trends are seen in many of the basins throughout northern and western China.

3.5.1   *Flora evidence for an Early Miocene monsoon*

The overall changes in vegetation and climate in China during the Cenozoic were summarized by Sun and Wang (2005) and shown in Figure 3.30. Two contrasting patterns of climate zone distribution can be recognized. During the Paleogene, an arid belt stretched across the country (Figure 3.30 (c)), while during the Neogene, aridity was focussed in the north-west (Figure 3.30 (b)). The Neogene situation is relatively close to the modern climatic zonation of China. Given the dominance of the Asian monsoon system in controlling climate in all parts of modern China, with summer monsoon rains dominating in the east and south, and winter monsoon winds in the north and west, Sun and Wang (2005) concluded that the pollen and facies constraints favored an initial

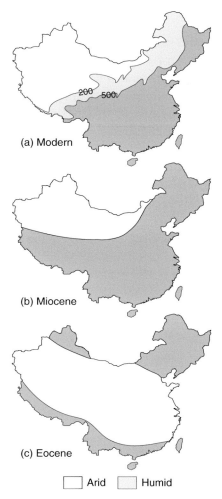

Figure 3.30 Distribution of arid and humid zones in China from Sun and Wang (2005).
(a) Modern; the 200 mm and 500 mm isohyets delineate the arid–semiarid zone
(Editorial Board of *Natural Geography of China*, 1979); (b) Miocene as an example of the
Neogene climate pattern; (c) Eocene exemplifying the Paleogene. Arid areas are shown
in white, and humid in gray. Reprinted with permission of Elsevier B.V.

intensification of the Asian monsoon around the Oligocene–Miocene boundary
(i.e., ~24 Ma). Such a conclusion is consistent with data such as the weathering
records from the South China Sea (Figure 3.14), from the eolian record in
the North Pacific (Rea *et al.*, 1998) and the onset of loess sedimentation dated
by Guo *et al.* (2002), but seems at odds with the 8 Ma intensification favored from
the Arabian Sea upwelling records.

3.5.2   Evidence from marine carbon

Despite this apparent contradiction between East and South Asian monsoons, an earlier monsoon intensification is consistent with carbon isotope data from the South China Sea. Black carbon generated by forest or bush fires on land is found in most oceanic pelagic and hemipelagic sediments. Like organic matter in paleosols the carbon isotope character of this marine black carbon reflects the C3/C4 ratios of the vegetation in the source region and can thus be used to reconstruct changing flora and climate onshore (Bird and Cali, 1998). Jia *et al.* (2003) analyzed the $\delta^{13}C$ values of black carbon in sediments deposited on the northern margin of the South China Sea (ODP Site 1148) in order to assess the evolving character of vegetation in southern China since 30 Ma. The results of this analysis are shown in Figure 3.31.

Carbon data from ODP Site 1148 are believed to represent the terrestrial vegetation, not marine plants, because the C/N ratios are much higher than the <10 typical for marine sources. The range of $\delta^{13}C$ values shows a progressive

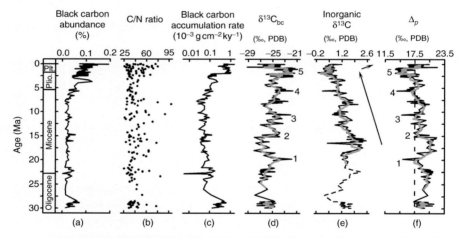

Figure 3.31 Analytical data from Ocean Drilling Program Sites 1147 and 1148 from Jia *et al.* (2003), reproduced with permission from the Geological Society of America. (a) Black carbon abundance in dry bulk sediments. (b) C/N elemental ratios in treated samples containing black carbon. (c) Black carbon accumulation rates. (d) Isotopic composition of black carbon. (e) Isotopic composition of marine inorganic carbon. The solid line represents planktonic foraminifer; the dashed line at the lower part of the curve is for bulk carbonate. (f) Difference between atmospheric $\delta^{13}C$ and $\delta^{13}C_{bc}$ as explained in the text. Thick gray lines in (d), (e), and (f) are 5-point running averages. Note the 2 My hiatus in late Oligocene. PDB – Peedee belemnite; Plio – Pliocene; Plt – Pleistocene. Numbers 1–5 show five isotope excursions from early Miocene to Pliocene–early Pleistocene.

evolution since 30 Ma, but with five marked, more positive, peaks in δ^{13}C values that might correlate with the expansion of C4 plants. Jia *et al.* (2003) chose to plot a factor, Δp, that is a measure of the difference between the measured δ^{13}C values of the black organic carbon and atmospheric carbon at the time of deposition because this is the base line against which the plant isotope character can be assessed and this changes through geologic time. The atmospheric carbon value was obtained from the analysis of δ^{13}C from the calcareous shell material preserved within the same sedimentary intervals. At ODP Site 1148 Δp values fall largely within the typical range for C3 plants but become less positive up section, i.e., more C4-like. Below a threshold of +17.5, Δp values were interpreted to represent strong C4 plant influence. Negative values in Δp typify C3 ecosystems and are usually related to moisture deficits (Conte and Weber, 2002). In contrast, C4 photosynthesis is commonly associated with hot, dry environments with warm-season precipitation in a low atmospheric pCO_2 background (Pearson and Palmer, 2000; Huang *et al.*, 2001). As a result, Jia *et al.* (2003) interpreted the South China Sea carbon record to indicate development of the East Asian flora and climate toward the modern monsoonal system. Because the earliest peak of δ^{13}C appeared about 20 Ma, the record was interpreted to indicate Early Miocene initiation of the monsoon system.

3.6   History of Western Pacific Warm Pool and the Monsoon

Although the Western Pacific Warm Pool is not a product of, or the principal trigger for, the Asian monsoon system it is the largest oceanographic phenomenon in the Western Pacific Ocean. By controlling the temperature of the water offshore of East Asia, the Warm Pool indirectly controls the temperature difference between the Pacific Ocean and continental Asia, and thus the intensity of the monsoon. The Western Pacific Warm Pool is partially a product of collision between Indonesia and Australia (Figure 3.32), preventing the flow of warm water into the Indian Ocean, and instead allowing it to pond east of the Philippines. Prior to the closure of the Indonesian gateway, the flow of warm water into the Indian Ocean must also have affected the intensity of the South Asian monsoon, by intensifying the winter and reducing the summer monsoon. The flow of warm water from the Pacific to the Indian Ocean has been impeded since the Miocene, though the degree of constriction must have been considerably greater during the Last Glacial Maximum when the sea level was much lower. During the Miocene, closure of the gateway may even have had global effects.

Plate tectonic reconstructions that place initial severing of the deep-water flow around 25 Ma (e.g., Hall, 2002; Kuhnt *et al.*, 2004) raise the possibility that

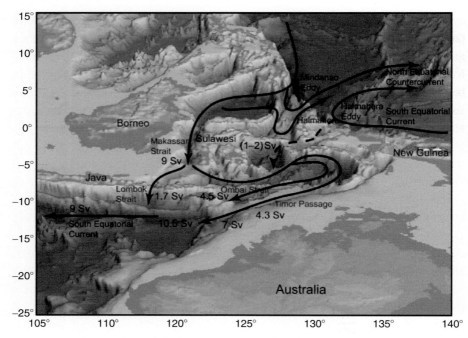

Figure 3.32 Present day Indonesian Throughflow, modified from Gordon (2001) and Gordon *et al.* (2003), and drawn by Kuhnt *et al.* (2004) on a topographic base-map derived from the ETOPO-5 bathymetric data set. Reprinted with permission of the American Geophysical Union.

the re-organization of currents triggered some of the paleoclimatic events recognized at this time. Miller *et al.* (1991) identified a major carbon isotope excursion at the base of the Miocene, while oxygen isotopes show first a warming around 24–26 Ma followed by a dramatic cooling after 23 Ma that might be linked with the collision (Zachos *et al.*, 2001). As described above, there is now significant evidence to suggest that initial monsoon intensification might also date from this time period.

By Early Miocene time, the distinct vertical $\delta^{18}O$ gradient in Indian Ocean waters, which was characteristic of the Oligocene, had weakened, as intermediate and deep waters warmed and cool deep-water masses remained restricted to the southern part of the Indian Ocean. Woodruff and Savin (1989) used the large observed $\delta^{13}C$ differences in benthic foraminifers to argue that the Indian Ocean was isolated from the Pacific Ocean in the Early and Middle Miocene. However, the Indian Ocean was in a state of strong convective over-turn. Young, $\delta^{13}C$-rich intermediate waters (1.0–2.5 km) originated in the northern Indian Ocean, but the southern and eastern Indian Ocean appear to have mainly been influenced by deep water of Antarctic origin.

Figure 3.33 shows oxygen and carbon isotope data compiled by Kuhnt *et al.* (2004) to show the differences that have existed on either side of the Indonesian gateway since ∼24 Ma. The chart shows the significant divergence in deep-water $\delta^{18}O$ values prior to 24 Ma. While tropical Pacific records, including those from the South China Sea (Zhao *et al.*, 2001a,b), indicate warming of deep waters coinciding with the major global carbon isotope excursion at the Oligocene–Miocene boundary (Miller *et al.*, 1991), the Indian Ocean records from ODP Sites

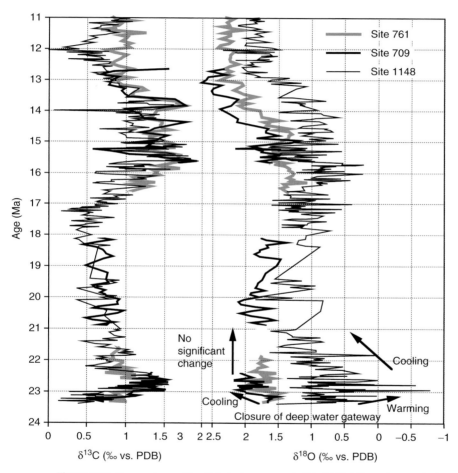

Figure 3.33 Comparison of benthic isotope records (11–24 Ma) from the tropical western Indian Ocean (ODP Site 709, (Woodruff *et al.*, 1990; Woodruff and Savin, 1991)), eastern Indian Ocean (Site 761, (Holbourn *et al.*, 2004)) and South China Sea (Site 1148, (Zhao *et al.*, 2001a,b)). All records were tuned to the age model of Site 761 using the carbon isotope record (Holbourn *et al.*, 2004).

709 and 761 indicate cooling at the same time. Cooling occurred later in the South China Sea (21–23 Ma; Kuhnt *et al.*, 2004), at a time when Indian Ocean deep water already had reached high $\delta^{18}O$ values. These data argue convincingly for an early deep-water closure of the Indonesian Flow-Through, and open up the possibility that severing of this flow could be a powerful influence on the intensification of the monsoon. Closure is fundamental to the formation of the Western Pacific Warm Pool (Godfrey, 1996), although its full development is younger and requires further closure of the gateway to take place before reaching its modern intensity (Cane and Molnar, 2001). Foraminiferal population and stable isotope data from ODP Site 806 on the Ontong Java Plateau (Nathan and Leckie, 2004) indicate there is compelling evidence for a proto-warm pool by at least 11.5–10 Ma. The presence of a warm pool implies tectonically driven constriction through the Indonesian Seaway, which needs to be accounted for when explaining the development of the Asian monsoon.

3.7   Summary

The long-term evolution of the Asian monsoon is still very much a topic of research, hampered by incomplete long-term sedimentary records from both East and South Asia, as well as some disagreement about what constitutes a robust monsoon proxy. As shown in Chapter 4, much of the monsoon research completed to date has focussed on orbital timescales, with less attention paid to the tectonic timescale. In Figure 3.34 we plot some of the more complete and robust climate proxies, in order to assess evidence for long-term monsoon evolution. It is noteworthy that no monsoon records extend beyond 17 Ma in South Asia. In the Arabian Sea paleoceanographic records only extend to 12 Ma. Although it is still debated whether the 8 Ma event first identified by Kroon *et al.* (1991) truly represents a strengthening of the summer monsoon driven by Tibetan surface uplift, it is clear that this was a time of oceanographic change and increased productivity. Dust in the North Pacific became coarser grained at this time and accumulation rates on the Loess Plateau accelerated (Rea, 1994; Ding *et al.*, 2001) showing that the winter monsoon winds in East Asia strengthened at this time. Similarly, the degree of chemical weathering in South Asian flood plains increased, although it seems likely that this reflects drying of the climate, not increased rainfall (Derry and France-Lanord, 1996, 1997; Dettman *et al.*, 2001). Falling clastic sediment flux in East Asia, as well as the Indian Ocean, is consistent with less precipitation starting around 8 Ma (Burbank *et al.*, 1993; Clift, 2006). Further drying and wind strengthening, at least in Central Asia, is clearly shown by both loess and oceanic eolian records around 3.0–2.5 Ma. Again, this change can be tied to global climatic change

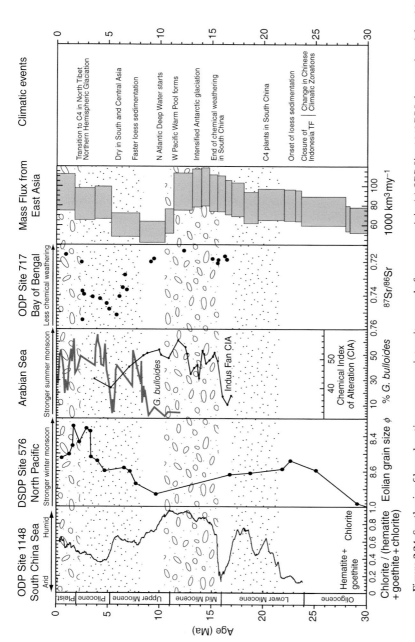

Figure 3.34 Synthesis of long-duration monsoon intensity records from across Asia. (a) ODP Site 1148 DRS data analyzed for chlorite content vs. hematite and goethite (unpublished data). (b) Grain size of eolian sediments from DSDP Site 576 in the North Pacific, from Rea (1994). (c) Abundance of *G. bulloides* from ODP Site 727 on the Oman margin, from Kroon *et al.* (1991), together with the Chemical Index of Alteration (CIA; Nesbitt and Young, 1982) from petreolum well Indus Marine A-1 on the Pakistan Shelf (unpublished data). (d) Sr isotope data from clay samples from ODP Site 717 in the Bay of Bengal, from Derry and France-Lanord (1996). (e) Mass flux budget for East Asian deltas is from Clift *et al.* (2004). Stippled patterns indicate strong rains and faster erosion, while the conglomerate motif indicates maxima in monsoon intensity.

events and, while representing milestones in the development of the Asian monsoon, is not necessarily linked to tectonic processes in Tibet.

It is noteworthy that the weathering record reconstructed by the Chemical Index of Alteration (CIA; Nesbitt and Young, 1982) in the Indus delta region tracks the general form of the weathering records from South China Sea, suggesting that the South and East Asian monsoon may evolve in parallel with one another over tectonic timescales. What is less clear is how these records correlate with the weathering records in the Bay of Bengal because the changes appear to be offset from one another. Nonetheless, even here there is a long-scale transition from wetter climates prior to ~10–8 Ma, followed by a drier period until 2–3 Ma, and finally renewed stronger monsoon since that time.

Initial intensification of the monsoon now appears to date from close to the Miocene–Oligocene boundary (24–22 Ma). This date is based on the change in clay mineralogy and carbon isotopes in black carbon in the South China Sea (Clift, 2006; Jia et al., 2003), the coarsening of eolian dust in the North Pacific (Rea, 1994), the onset of loess accumulation in North China (Guo et al., 2002) and the large-scale re-organization of climatic zones in China (Sun and Wang, 2005). All of these are set against a background of increasing clastic sediment flux to the marginal seas of Asia (Clift et al., 2004) that cannot yet be dated well enough to show a clear linkage.

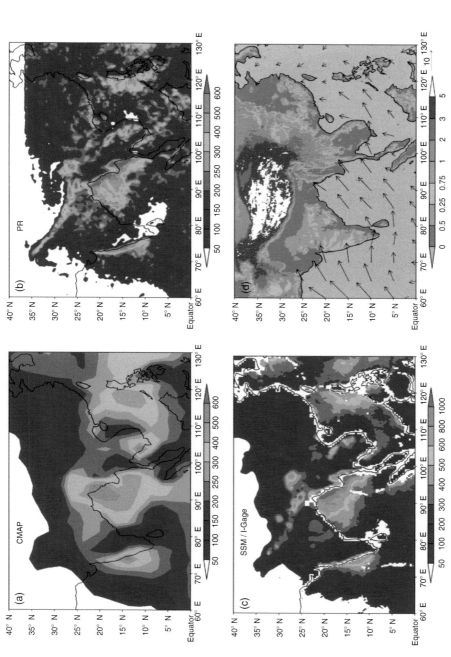

Figure 1.16 Summertime rainfall over the northern Indian Ocean and South Asia. Frames (a)–(c) show June–August mean rainfall rates (mm per month) from different data sources: (a) the Climate Prediction Center Merged Analysis of Precipitation (CMAP; 2.5° resolution, coverage 1979–2003); (b) high resolution data from the space-based Tropical Rainfall Measurement Mission Precipitation Radar (1998–2004) and (c) merged high resolution data from the space-based Special Sensor Microwave Imager and land-based observations (1987–2003). Surface topography (km) is shown in frame (d). (Xie *et al.*, 2006).

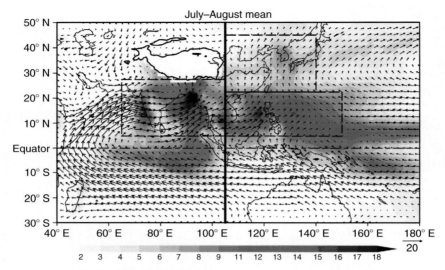

Figure 2.1 Climatological July–August mean precipitation rates (shading in mm day^{-1}) and 925 hPa wind vectors (arrows) from Wang *et al.* (2003a). The precipitation and wind climatology are derived from CMAP (Xie and Arkin, 1997) (1979–2000) and NCEP/NCAR reanalysis (1951–2000), respectively. The three boxes define major summer precipitation areas of the Indian tropical monsoon (5–27.5° N, 65–105° E), WNP tropical monsoon (5–22.5° N, 105–150° E) and the East Asian subtropical monsoon (22.5–45° N, 10–140° E). Reprinted with permission of Elsevier B.V.

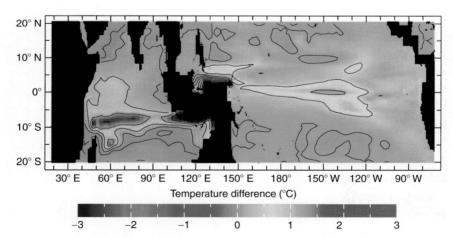

Figure 2.12 The difference in temperatures at 100 m depth between two ocean GCM runs from Cane and Molnar (2001). The sole difference between conditions for the runs is that in one the northern tip of New Guinea–Halmahera is at 28° N, and in the other it is at 38° S. Cane and Molnar suggest that the pattern showing the difference of the former minus the latter approximates the difference between that at present minus that at 4 Ma. Reproduced with permission of Macmillan Magazines Ltd.

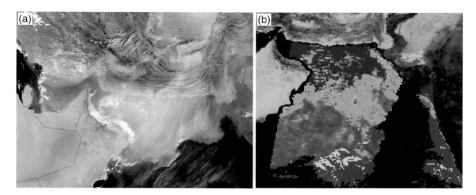

Figure 3.1 (a) A MODIS image from NASA (Visible Earth) of the Arabian Sea during the summer monsoon showing a storm blowing dust from Arabia to the north-east across the Gulf of Oman toward Pakistan. (b) A false color image from NASA showing chlorophyll in surface waters during the winter monsoon. Wind blowing south is driving upwelling and enhanced productivity along the Makran coast. Images reproduced with permission of NASA.

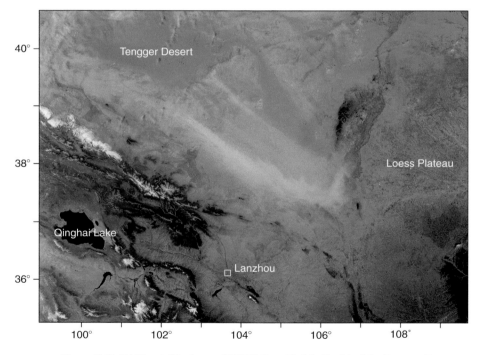

Figure 3.15 NASA satellite image (MODIS from Visible Earth) of the Tengger Desert in north-west China showing a powerful winter dust storm putting large volumes of material into the atmosphere, which is being transported to the south-east. Note: Qinghai Lake and the upper reaches of the Yellow River are in the bottom left hand corner. Images reproduced with permission of NASA.

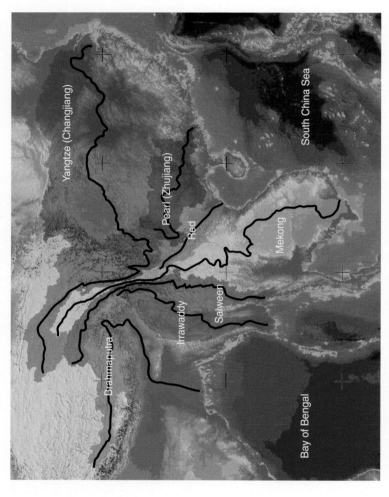

Figure 5.7 Shaded topographic relief map of East Asia showing the modern drainage basins of the largest rivers that bring material from eastern Tibet to the marginal seas. Note that the river courses all run close together in eastern Tibet, close to the eastern syntaxis of the Himalaya, making drainage capture from one system to another relatively simple. Figure modified from Clark *et al.* (2004). Reproduced with permission of the American Geophysical Union.

4

Monsoon evolution on orbital timescales

4.1 Introduction

The Asian monsoon system varies not just over long periods of geologic time (10^6–10^7 years), but also on millennial and shorter timescales. In this chapter we explore what is known about this shorter-term development in monsoon intensity and explain some of the processes that control this. Since the onset of icehouse conditions around three million years ago the global climate has largely been linked to, and possibly controlled by, the changing intensity of Northern Hemispheric Glaciation. This, in turn, is now well established as being largely a function of solar energy variations, driven by long-term fluctuations in the Earth's orbit around the Sun (e.g., Hays *et al.*, 1976; Shackleton and Opdyke, 1977; Martinson *et al.*, 1987; Shackleton *et al.*, 1990; Bassinot *et al.*, 1994). Although the link between the growth and retreat of glaciers in the northern hemisphere and the intensity of the Asian monsoon might not be immediately apparent, it has been recognized that major regional climatic phenomena do appear to influence each other through a series of teleconnections. Figure 4.1 shows the proposed connections between different regional oceanographic and climatic systems, in an attempt to demonstrate how changes in one system can feedback and affect climate in other parts of the planet. The two most important teleconnections we shall consider here are those between the El Niño Southern Oscillation (ENSO) system of the Pacific Ocean and the intensity of Northern Hemispheric Glaciation. The ENSO control on monsoon intensity occurs on relatively short timescales and is discussed in greater detail in Chapters 1 and 6. In this chapter we consider the changes in monsoon intensity over orbital timescales, mostly during the Pleistocene and Holocene, where the superior age control allows short-term fluctuations to be documented and quantified.

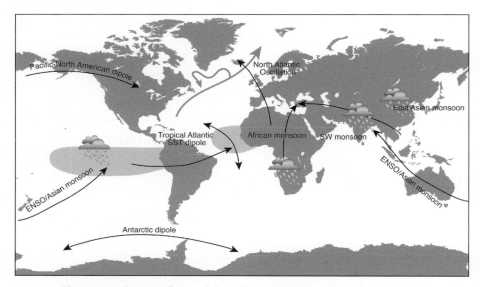

Figure 4.1 Schematic figure of the major components of the global climate system, re-drawn and modified from Zahn (2003). The figure shows some of the major linkages between climate systems operating between different ocean basins. Changes in one or another of these can result in a cascading effect that causes changes in the strength of the Asian monsoon on a variety of timescales. Reproduced with permission of Macmillan Magazines Ltd.

4.2   Orbital controls on monsoon strength

The intensity of the Asian monsoon (and much of the global climate) is tied to the delivery of energy from the Sun, which in turn varies over geologic timescales. As described in this chapter, the Asian monsoon system has varied greatly in its intensity over the past series of glacial–interglacial cycles (Van Campo *et al.*, 1982; Clemens *et al.*, 1991; Sirocko *et al.*, 1993; Overpeck *et al.*, 1996a). Evidence from both marine sediments and modeling results suggests that variations in the monsoon are a response to changes in Northern Hemisphere insolation or to glacial boundary conditions (Prell and Kutzbach, 1987; Clemens *et al.*, 1991, 1996; Sirocko *et al.*, 1993; Overpeck *et al.*, 1996a). The Earth's orbit shows three types of long-term variation that have had an important effect on the global climate system. Figure 4.2 shows how the eccentricity of the planet's orbit can affect the delivery of solar energy owing to the changing shape of the orbit and thus the distance from Sun. This variability changes over timescales of ~100 ky.

The tilt of the axis of Earth's orbit, known as the obliquity, varies from 24.5° to 22.1° over the course of its 41 000 year cycle. The current angle of axial tilt is 23.4°.

Eccentricity, 100 000 y

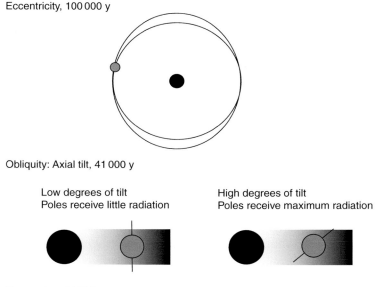

Obliquity: Axial tilt, 41 000 y

Low degrees of tilt
Poles receive little radiation

High degrees of tilt
Poles receive maximum radiation

Precession, 21 000 y

Today: perihelion during northern winter

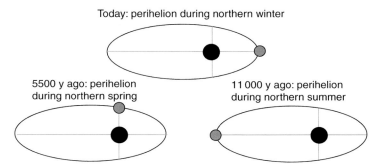

5500 y ago: perihelion
during northern spring

11 000 y ago: perihelion
during northern summer

Figure 4.2 Figures showing the three major styles of irregularity in the Earth's orbit that drive changes in global climate, including the Asian monsoon.

Changes in axial tilt affect the distribution of solar radiation received at the Earth's surface. When the angle of tilt is low, polar regions receive less solar energy (insolation). When the tilt is greater, the polar regions receive more insolation during the course of a year. Like precession and eccentricity, changes in tilt thus influence the relative strength of the seasons, but the effects of the tilt cycle are particularly pronounced in the high latitudes where Northern Hemispheric Glaciation began. Ice formed during the polar winter will tend to melt during summers when tilt is great, thus making the establishment of permanent ice caps more difficult. Conversely, if the tilt acts to minimize heat delivery during the summer of either hemisphere then this can increase the building of permanent ice fields and aid the process of widespread continental glaciation. The third form of orbital variation is precession. Like a spinning top, the Earth's orbit wobbles so that over the course of a precessional cycle, the Earth's rotation axis traces a circle in space. This wobble causes the precession of the equinoxes. The cycle of precession takes about 21 000 years to complete (Figure 4.2). Precession influences the delivery of heat to the high latitudes and affects the ability of permanent ice sheets to form and grow.

The relationship between monsoon strength and orbital irregularities was first investigated on the western margin of the Arabian Sea offshore of Oman. Clemens and Prell (1990) investigated the accumulation of wind-blown dust at Core Site RC27-61 (Figures 4.3 and 3.3) in order to reconstruct the varying strength of winds and the changing aridity of Arabia through time. Sediment trap data (Figure 4.4) show that there is a strong association between strong sustained monsoon winds of the summer offshore Oman, the rate of accumulation of eolian dust and the foraminifer *G. bulloides* that is associated with monsoon-driven upwelling. The size of the dust particles is more closely linked to the carrying capacity and speed of the wind, while the total accumulation rate is controlled by the total supply of dust that is controlled by the aridity of Arabia.

Clemens and Prell (1990) were able to show a dramatically changing mass accumulation rate in eolian dust during the last 400 ky, which correlated closely with the changing $\delta^{18}O$ values measured from the calcareous shells of planktonic foraminifers in the same core. Similarly, Krissek and Clemens (1991) showed that the mineralogy of the eolian dust changed over orbital timescales during the Pleistocene, reflecting changing sources and wind strengths. The $\delta^{18}O$ values of shell material reflect the value of the sea water at the time of precipitation and this in turn is controlled by the temperature and salinity of the sea water. As a result, $\delta^{18}O$ values from foraminifers have been extensively used to track glacial–interglacial climate cycles because sea water is colder during glacial times.

The close correspondence between eolian mass accumulation rate and $\delta^{18}O$ values indicated that the aridity of Arabia was controlled over millennial timescales by the same processes that caused Northern Hemispheric Glaciation during the Late Pleistocene, with drier periods in Arabia corresponding to glacial periods. A similar conclusion was reached in the same region by Sirocko *et al.* (2000) on the basis of geochemical data from the last 25 ky. A series of chemical proxies was used to reconstruct eolian dust flux into the Arabian Sea since the Last Glacial Maximum and these showed a strong link with glacial periods (Reichart *et al.*, 1997). Figure 4.5b shows the rapidly changing grain size of the eolian dust deposited since 400 ka offshore of Oman. These data testify to changing summer monsoon wind strengths over relatively short periods of

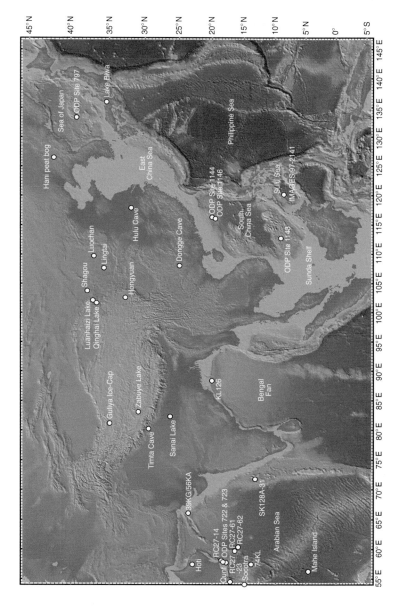

Figure 4.3 Shaded topographic and bathymetric map of the Asia–Pacific region showing the location of the major study sites considered in this chapter.

99

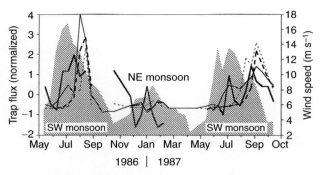

Figure 4.4 Sediment trap fluxes associated with summer and winter monsoon circulation from Clemens *et al.* (1996). Trap data are normalized for purposes of plotting. Average inputs are shown for peak winter (December through March) and summer (June through September) periods, respectively: *G. bulloides* (light dashed line), 369 and 2071 individuals m^{-2} day^{-1}; lithogenic flux (light solid line), 16 and 48 mg m^{-2} day^{-1}; lithogenic grain size (median diameter, volumetric distribution) (heavy solid line), 10.9 and 12.6 μm; and opal flux (heavy dashed line), 13 and 55 mg m^{-2} day^{-1}. Wind speed (hatched pattern) is shown on the right axis. Reproduced with permission of the American Association for the Advancement of Science.

geologic time. Clemens and Prell (1990) compared the grain size pattern with the changing insolation driven by the precessional cycle and noted the close comparison between the two, a link confirmed by recent work on nanofossil assemblages from the same region (Rogalla and Andruleit, 2005). This diagram provided some of the best initial evidence that monsoon strength was linked to glacial cycles and, in turn, to orbital processes over millennial timescales.

One complexity of this comparison was that Clemens and Prell (1990) were only able to make the grain size record and the precessional cycle match after shifting the precessional record by 9 ky, an observation that implies a lag or lead in the relationships between the two. This relationship can be understood more fully using a phase wheel diagram (Figure 4.6). The phase wheel is used to describe the lag between the forcing function (here the precessional component of the Milankovich orbital cycle) and the response that is usually traced with an oceanographic or climatic proxy. The phase difference between an orbital component and a given climatic record is plotted as a vector whose angle is measured clockwise (positive, lagging) or counterclockwise (negative, leading) from the zero point. The error associated with the phase estimates is shown by the shaded region centered on the vector. The zero point on each wheel was set by Clemens and Prell (1990) to be at maximum eccentricity, maximum obliquity and June 21 perihelion (Figure 4.2), which is useful because for the 41 000 y obliquity cycle this corresponds to maximum summer insolation.

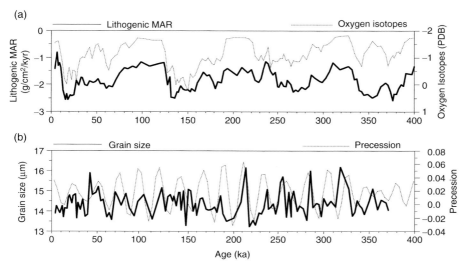

Figure 4.5 (a) The lithogenic mass accumulation rate (multiplied by -1) and the $\delta^{18}O$ records from core RC27-61 from Clemens and Prell (1990). These data show a positive correlation indicating that the flux of terrigenous material is related to the extent of global ice volume and the associated glacial environment and, by analogy, to external insolation forcing over Milankovitch timescales. (b) With respect to timing, the large-amplitude events in the lithogenic grain size record can generally be accounted for by maxima and minima in the Earth's orbital precession record (dashed line, phase lagged by 9 ky) indicating the influence of insolation forcing over Milankovitch timescales. Reprinted with permission of the American Geophysical Union.

All three orbital parameters (eccentricity, precession and obliquity) show a coherent development with eolian lithogenic mass accumulation rates and foraminifer $\delta^{18}O$ values. There is also a clear correspondence between eolian mass accumulation rates offshore of Oman and the magnetic susceptibility of Chinese loess (An *et al.*, 2001) that suggests a pan-Asian linkage. Loess–paleosol deposits show simple links between magnetic susceptibility and the monsoonally modulated intensity of chemical weathering. The similarity between the Arabian eolian record and the loess deposits indicates that climate change in Arabia and in central China are linked and presumably controlled by similar processes.

Surprisingly, grain size, interpreted as a proxy for summer monsoon wind strength in the Arabian Sea, only shows coherency with the precessional cycle, where the phase wheel shows a lag of $\sim 148°$, corresponding to around 9 ky. Although grain size lags the precessional forcing it is in phase with maxima in *G. bulloides* abundances and in sea-surface temperatures, as might be anticipated

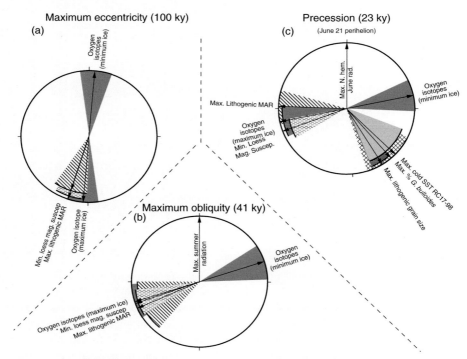

Figure 4.6 Phase wheels for (a) the eccentricity, (b) obliquity and (c) precession bands from Clemens and Prell (1990). All records shown are statistically coherent with the orbital forcing and with one another, within a given cluster, with (1) the *G. bulloides* record from core PC17-98 of sea-surface temperature (SST), and (2) the loess susceptibility record, which is not significantly coherent with the core PC27-61 lithogenic mass accumulation rate record though they share a near-zero phase relationship. The error associated with the phase estimates is shown by the shaded region centered on the vector. Reprinted with permission of the American Geophysical Union.

given that all three are linked to the intensity of summer monsoon winds. It is noteworthy that Arabian Sea eolian mass accumulation rates are not in phase with the grain size variability, similar to the situation found in the North Pacific (Rea *et al.*, 1998). This observation confirms that these materials are deposited by wind, not run-off from the continent, and that the two are controlled by different processes (i.e., aridity and monsoon wind strength). The lack of a strong linkage between the 100 ky eccentricity cycle and the grain size indicates that the relationship between summer monsoon strength and North Hemispheric glaciation is not straightforward. Glacial intensity has largely been controlled by the eccentricity cycle during the Pleistocene.

4.3   Eolian records in North-east Asia

4.3.1   *Marine records in the Sea of Japan*

High-resolution eolian marine sediment records spanning the Late Pleistocene are also known from East Asia, especially from the Sea of Japan. The Sea of Japan is largely isolated from the Pacific Ocean (Figure 4.7) and during glacial times its paleoceanography is strongly controlled by changing sea levels. In an attempt to reconstruct the changing climate of the region Tada *et al.* (1999) described centimeter to meter-scale alternations of dark and light colored layers in the sediment that are found across the entire basin where it was drilled by Ocean Drilling Program (Figure 4.8). These workers proposed that during inter-glacial periods the Sea of Japan became stagnant owing to a more humid climate with a stronger East Asian summer monsoon that caused increased fluvial water run-off from the Yangtze and Yellow Rivers. This less saline, but nutrient-rich water was delivered to the Sea of Japan by the East China Sea Coastal Water, a north-flowing oceanographic current, that strengthened shallow-water biogenic production, while simultaneously reducing deep-water circulation. A similar oceanographic effect, driven by monsoon-derived discharge from the Amur River, was proposed for the Sea of Okhotsk by Harada *et al.* (2006). Figure 4.8 shows that the variability in organic carbon content, oxygenation of the sea floor (measured by the degree of bioturbation) and diatom abundance (reflecting productivity) all show strong coherent variability and that they correlate with the marine isotopic stages that define the alternating glacial–interglacial climate of the northern hemisphere. Thus the paleoceanography of the Sea of Japan, which is known to be heavily influenced by both summer and winter monsoons, is seen to vary on a millennial timescale that correlates with orbitally forced glacial climate and extends back at least to 200 ka.

As well as river sediments the Sea of Japan also receives eolian dust from Central Asia, mostly blown eastwards into the Pacific Ocean by winter monsoon winds (Rea, 1994). Known as loess in China this material is called "Kosa" in Japan, meaning "yellow sand." Tada *et al.* (1999) noted coherency between the other oceanographic indicators and the relative proportion of Kosa material in the coarser-grained clastic component of the sediment. The ODP Site 797 is located on the flanks of the Yamato Rise (Figure 4.7) and its sedimentary cover can be used to examine the strength of eolian sedimentation because the sediments found here cannot have been influenced by continental sources either from Honshu or Korea. Irino and Tada (2002) used a combination of major element chemical data and mineralogy to define four populations in the clastic component of the sediments drilled at ODP Site 797. These are typical loess, weathered loess, arc-derived mudstones from Japan and Quaternary tephras. "Weathered" loess is

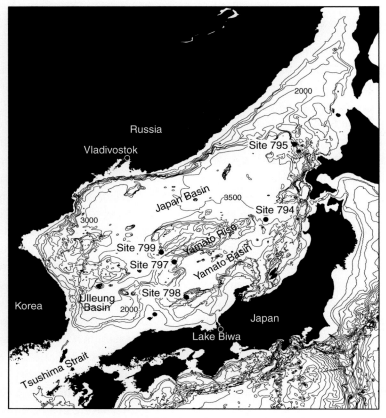

Figure 4.7 Bathymetric map of the Sea of Japan showing the location of scientific boreholes and principal geographic features named in the text. Water depths in meters from GEBCO compilation.

distributed in China within the range of the summer monsoon and was originally defined as "Relict Carbonate Loess" by Gong *et al.* (1987). The heavier rains of the summer monsoon allow this loess to experience chemical weathering, while that preserved to the north remains dry and relatively fresh.

The composition of Kosa reflects the mixture of "typical" loess and "weathered" loess, which suggests that Kosa is supplied not only by arid Chinese inland areas but also by peripheral soil. Figure 4.9 shows the temporal variation in the Kosa and arc flux since ~200 ka (Marine Isotope Stage 7). Both Kosa and arc debris appear to be more abundant in marine sediments during periods of glacial maximum, especially Stage 2 (~20 ka) and Substage 6.2 (~140 ka). The threefold increase seen in the Kosa constribution could be accounted for by a twofold increase in the area of the loess source region during glacial times. Although theoretically a fall in sea level during glacial times could deliver loess-rich

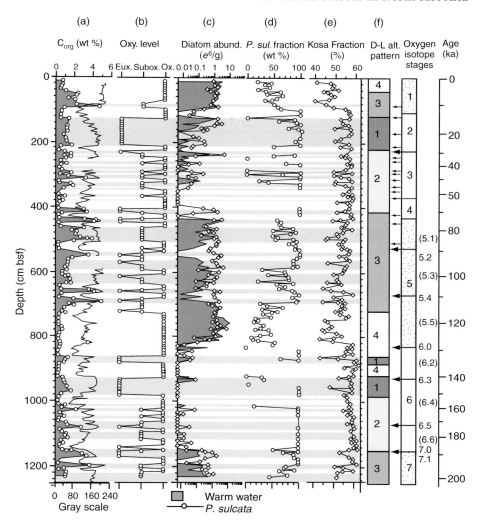

Figure 4.8 Study of Tada *et al.* (1999) showing profiles of (a) C_{org} and gray-scale,
(b) bottom water oxygenation level, (c) warm water diatom and *Paralia sulcata*
abundance, (d) *P. sulcata* fraction, (e) Kosa fraction and (f) patterns of alternation
of the dark and light layers at ODP Site 797. Dark layers are shaded. Correlation
with the oxygen isotope stages and substages are also shown. Dark layers
correlated with KH-79-3 to construct the age model are indicated with smaller
arrows whereas oxygen isotope stage tentatively identified on the basis of the
warm water diatom abundance curve and two marker tephra layers are
indicated by larger arrows. Reprinted with permission of the American
Geophysical Union.

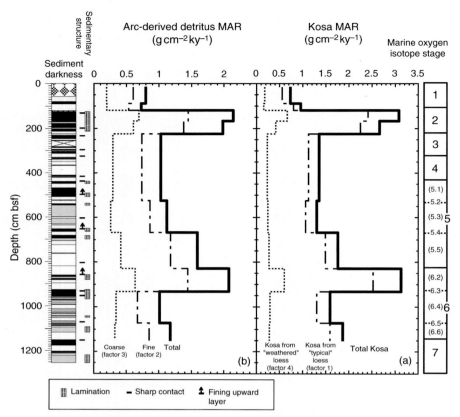

Figure 4.9 Diagram from Irino and Tada (2002) showing the temporal variations in (a) Kosa and (b) arc-derived detritus mass accumulation rate at ODP Site 797. Positions of dark layers and primary sedimentary structures are shown at the bottom. Oxygen isotope stages are based on Tada *et al.* (1999). MAR = mass accumulation rate. Reprinted with permission of Elsevier B.V.

sediment direct from the Yellow River to the Sea of Japan through the Tsushima Straits Saito (1998) has argued that because precipitation is believed to be less during glacial periods the run-off in the Yellow River would be much less and thus that the extra Kosa sedimentation must reflect delivery by a strengthened winter monsoon wind during glacial times. This process is further enhanced by increased aridity and dust production in Central Asia brought on by a weaker summer monsoon.

Decreasing Kosa flux into the Sea of Japan most likely reflects reduced availability of "typical" loess due to expansion of high soil moisture area. An increase in arc-derived sedimentation during glacial periods is more surprising because precipitation might have been expected to decrease in Japan at this time and that in turn might be expected to drive less, not more, erosion. The increase might instead reflect a more efficient transport across a stratified basin at the time of deposition of the dark organic-rich layers (Tada *et al.*, 1999).

The work on the monsoon around the Sea of Japan represents an important development in our understanding of the monsoon gained from the Arabian Sea. The relative contributions from typical and weathered loess are seen to vary significantly between glacial–interglacial cycles. Kosa from "weathered" loess shows fluctuation with approximately 40 ky periodicity (i.e., obliquity, not precession) with higher contributions seen during around Marine Isotope Stages 2 and 4 and Substages 5.4 and 6.3 (12–30 ka, 62–74 ka, ~110 ka and ~140 ka; Irino and Tada, 2000, 2002). The fact that such a 40 ky periodicity is not observed in the variation of Kosa from "typical" loess indicates that another mechanism controls the erosion and delivery of "weathered" loess to the Sea of Japan. "Weathered" loess is distributed to the south of Yellow River, so that if the main storm track shifted southward "weathered" loess would be more efficiently eroded and potentially transported to the Sea of Japan. The location of the average storm track is closely linked to the intensity of the Siberian atmospheric high during winter and spring time (Pye and Zhou, 1989). The importance of the obliquity orbital component in controlling the flux of "weathered" loess was an important discovery, yet does not diminish the dominant 21 ky precessional control on monsoon intensity in East, as well as South Asia during the Late Pleistocene.

4.3.2   *Lake records in Japan*

Quaternary lakes in Japan might be expected to provide relatively high resolution, undisturbed climatic records of monsoon variability on orbital timescales, similar to those found in the Sea of Japan. In particular, Lake Biwa, which is the largest lake in Japan and is located in central Honshu (Figure 4.7) has been the focus of coring and climate studies because it contains a sediment record that is known to span 130 000 y, based on the occurrence of distinctive volcanic tephra layers found in the section (Yoshikawa and Inouchi, 1991, 1993). In a study by Xiao *et al.* (1999) the strength of the winter and summer monsoon in central Japan was reconstructed using the clastic lake sediments. Sedimentary grains >20 μm across were presumed to have been washed into the lake by fluvial run-off because precipitation is the strongest control on slope erosion around the lake. In contrast, those grains that are <10 μm across were interpreted as having been carried to the lake by winter monsoon winds (Xiao *et al.*, 1997). Figure 4.10 shows the result of the analysis, and in particular the inverse correlation between eolian quartz flux (EQF) and the fluvial quartz flux (FQF). This implies that at times when the summer monsoon was strong the winter monsoon was weak, and vice versa. The times of minimum eolian quartz flux

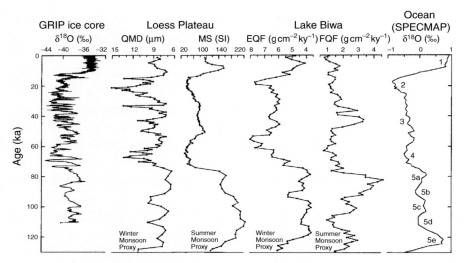

Figure 4.10 Monsoon proxy evolution since 130 Ka from Xiao *et al.* (1999), showing quartz median diameter (QMD, μm) and magnetic susceptibility (MS) of the Luochan loess–paleosol sequence and the eolian quartz flux (EQF, $g\,cm^{-2}\,ky^{-1}$) and fluvial quartz flux (FQF, $g\,cm^{-2}\,ky^{-1}$) of the Lake Biwa sediments with Greenland ice sheet $\delta^{18}O$ record (Dansgaard *et al.*, 1993) and SPECMAP $\delta^{18}O$ record (Martinson *et al.*, 1987). Lake Biwa EQF and FQF records are dated by interpolating between ages of the widespread volcanic ashes from Yoshikawa and Inouchi (1993). Reprinted with permission of Elsevier B.V.

were 125–73 ka and around 5.5 ka (maximum summer monsoon). In contrast, maximum eolian quartz flux occurs at 73–13 ka.

Comparison of the Biwa Lake record with similar aged sequences from the Loess Plateau shows a first-order similarity, suggestive of a common monsoon control to the sedimentation. In the loess, high magnetic susceptibility is typically associated with soil formation, and thus more humid, stronger summer monsoon conditions. Conversely, greater median diameters for quartz grains indicate stronger winter monsoon winds. These proxies correlate closely with those seen in Lake Biwa, and in turn these both correlate with major fluctuations in global ocean $\delta^{18}O$ values that define the glacial–interglacial cycles. Xiao *et al.* (1999) were able to show strong millennial-scale variability in monsoon strength linked to global climate change in the Late Pleistocene (Figure 4.10). Slight lags between the $\delta^{18}O$ values and the Biwa monsoon proxies during the transitions between Marine Isotope Stages 5 and 4 (∼75 ka), as well as Stages 2 and 1 (∼14 ka) may reflect the large distances between the sources of the eolian dust and Lake Biwa. During glacial times when the dry regions expanded there may have been a delay in delivery of dust to the lake as the deserts grew and expanded closer to Japan. In contrast, there was a more immediate response in the Loess Plateau because it is located closer to the source regions.

More recently, Kuwae *et al.* (2004) examined diatom populations from the same Lake Biwa core. Diatoms are planktonic microorganisms, whose presence can be diagnostic of specific climatic conditions. *Stephanodiscus suzukii* was chosen because it is a form that thrives in the presence of high phosphorus levels, which reflect, in turn, summer precipitation levels. In addition, *Aulacoseiva nipponica* abundance was documented because it records vertical mixing of lake water, which is induced by winter temperatures and snowfall levels. These factors are in turn largely controlled by the strength of the East Asian winter monsoon. Spectral analysis of the changing *Stephanodiscus suzukii* abundance since 140 ka shows statistically significant periodicities centered on approximately 1400–1500, 1050–1150, 950–1050, and 850 y. The 1400–1500 y periodicity, in particular, is close to well known Dansgaard–Oeschger cycles (Bond *et al.*, 1997) and the Holocene Bond cycle (Bond *et al.*, 2001), implying a link between past summer rainfall in East Asia and temperatures in the North Atlantic.

The time interval 69–13 ka at Lake Biwa shows many intervals with low diatom flux rates, which are interpreted as indicating low phosphorus levels, possibly due to decreased summer monsoon precipitation levels (Kuwae *et al.*, 2004). In contrast, periods when total diatom flux exceeded the present level were also found during the time interval 54–13 ka, which might imply periods of stronger summer monsoon, but may be triggered by local lake circulation processes. Between 13 and 7 ka, *S. suzukii* and total diatom abundance records show low levels that imply low phosphorus supply and thus reduced summer rainfall. This is surprising because this period immediately following deglaciation is generally reconstructed as one of increasing summer monsoon strength (e.g., Herzschuh, 2006). The low values may principally reflect the cold and dry conditions of the Younger Dryas (13–11 ka), while between 11 and 7 ka, low values are driven by the changing lake convection patterns. Between 7 and 0 ka, *S. suzukii* and *A. nipponica* records show abundance near or above those of the present, suggesting winter temperatures, snowfall and summer precipitation levels probably similar to or above those currently recorded, i.e., under the influence of a strong summer monsoon.

4.3.3   *Eolian records from the Loess Plateau of China*

The most complete continental monsoon records in East Asia lie in the Loess Plateau of northern China. These sequences are mostly dated by paleomagnetic methods, supplemented by ^{14}C and thermoluminesence methods (e.g., Heller and Liu, 1982; An *et al.*, 1991, 1994), which allow them to be correlated with marine sequences. Porter and An (1995) compared the temporal evolution of grain size in the Chinese loess with millennial-scale variations in North

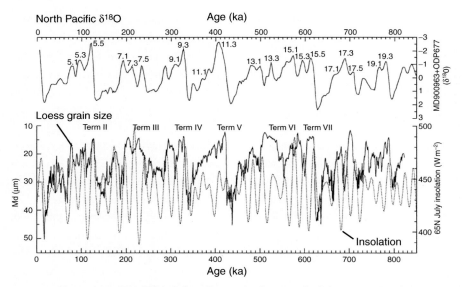

Figure 4.11 Comparison of median grain size record of Shagou section from Wu *et al.* (2005) (solid line, calculated using the >40 μm fraction), global ice volume (solid line, Bassinot *et al.*, 1994), and insolation in July at 65° N. Small numbers on upper plot of marine δ¹⁸O record show the Marine Isotopic Stages. Term = termination, end of glacial period. Reprinted with permission of Elsevier B.V.

Atlantic sea-surface temperatures and showed a close correlation between the two, indicative of large-scale teleconnections between these climate systems (Fang *et al.*, 1999). More recently Wu *et al.* (2005) examined an 830 ky record from Shagou in the western Loess Plateau (Figure 3.22). Like the earlier work, this study showed that the ages of the boundary between a paleosol and its underlying loess layer are very close to that of terminations to glacial periods recorded in the marine record in the North Atlantic. The Shagou section indicates that grain size changed sharply during each deglaciation, fining as warming occurred. This implied weaker winter monsoon winds as deglaciation progressed (Sun, 2004). Glacial terminations occurred at the time of the maximum July radiation in the Northern Hemisphere at high latitudes, consistent with the hypothesis that orbitally forced insolation is the primary factor affecting changes in monsoon intensity in the Chinese loess (Ding *et al.*, 1995). However, variations in the amplitude of the insolation and grain size records are not proportionally matched. For example, Termination II shows a weak response in grain size, while Termination VII is the most dramatic (Figure 4.11). As a result, the mechanism by which deglaciation and winter monsoon weakening are linked is not yet clear.

The different types of orbital forcing on Asian monsoon strength have been investigated by a number of studies of the loess. Liu *et al.* (1999) examined a

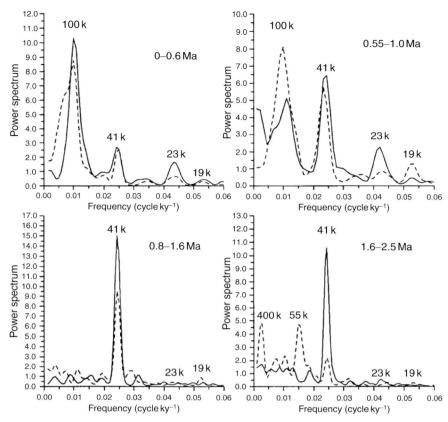

Figure 4.12 Comparison of the spectra of the Baoji loess grain size record (dashed line) with those of the DSDP Site 607 $\delta^{18}O$ record (solid line) over four different time intervals from Liu *et al.* (1999). Reprinted with permission of Elsevier B.V.

2.5 Ma record from Baoji in the southern plateau (Figure 3.22). In this place spectral analysis shows that prior to 1.6 Ma the loess differs from the marine $\delta^{18}O$ record of the North Pacific in showing no dominant frequency to its variability (Figure 4.12). This contrasts with the 41 ky forcing seen in the $\delta^{18}O$ record. However, more recently, Sun and An (2004) demonstrated a 410 ky, eccentricity-related cyclicity, in which each paleosol is equivalent to two interglacial oxygen isotope stages. It appears that 1.2 Ma was a key time for the monsoon because median grain size increased quickly after that time, signaling a stronger winter, and possibly summer, monsoon (Xiao and An, 1999). Moreover, a 41 ky (obliquity driven) cyclicity then entirely dominated the loess–paleosol alternations until ~1.0 Ma (Figure 4.12). After that time a 100 ky (eccentricity driven) cyclicity begins to become more important, so that by 0.5 Ma this is the major control on winter monsoon intensity. The discrepancy of climate cyclicity

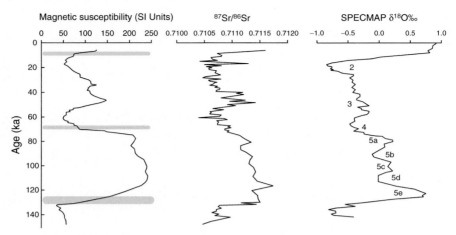

Figure 4.13 Time series comparing ⁸⁷Sr/⁸⁶Sr ratios of loess–paleosol carbonates, magnetic susceptibility (An *et al.*, 1991, 1994) in Luochuan section, and the SPECMAP stacked δ¹⁸O record of global ice volume (Martinson *et al.*, 1987). Numbers indicate Marine Isotope Stages (MIS). The diagram is modified from Yang *et al.* (2000). Reprinted with permission of Elsevier B.V.

between loess and deep-sea records over the 2.5–1.6 Ma interval suggests that the pre-1.6 Ma climate evolution cannot be understood simply by a regular 41 ky cycle model on a global scale. There appears to have been a watershed event at 1.6 Ma involving major monsoon strengthening, possibly driven by intensified Northern Hemispheric Glaciation.

Further complexities in the relationship between monsoon strength and orbital forcing have been revealed by a high-resolution study of loess at Luochuan (Figure 3.22). Yang *et al.* (2000) studied a 160 ky long record of magnetic susceptibility, which they compared with the global SPECMAP δ¹⁸O record for oceanic water compositions. Magnetic susceptibility is high in the paleosols developed under humid conditions of a strong summer monsoon, owing to the formation of magnetic minerals during diagenesis. Changes in susceptibility in the Luochuan loess section correlate closely with the SPECMAP record and indicate that the changes in summer monsoon strength in northern China are controlled by orbital processes.

The study of Yang *et al.* (2000) also measured Sr isotopes from the same section (Figure 4.13). Although Sr can be used as a provenance tool, these workers only analyzed secondary calcite formed after deposition in order to use this system as a proxy for the intensity of chemical weathering. They found that although the broadscale evolution of ⁸⁷Sr/⁸⁶Sr values paralleled the glacial–interglacial cycles, a series of high frequency, high amplitude fluctuations in ⁸⁷Sr/⁸⁶Sr values is superimposed on the last glacial–interglacial cycle since 130 ka. These short perturbations occurred on millennium timescales with an

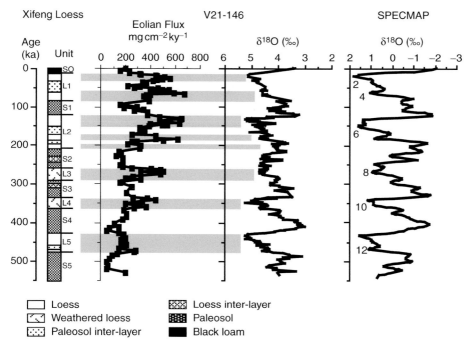

Xifeng Loess · Eolian Flux mg cm^{-2}ky^{-1} · V21-146 · δ^{18}O (‰) · SPECMAP · δ^{18}O (‰)

Age (ka) · Unit

Loess
Weathered loess
Paleosol inter-layer
Loess inter-layer
Paleosol
Black loam

Figure 4.14 Eolian flux and δ^{18}O data from the North Pacific after Rea (1994) showing correlation of dust fluxes to the loess stratigraphy at the Xifeng section in China and to the δ^{18}O record of global climate change. Dust flux maxima are of glacial age and correspond in time to the loess layers. Dust flux minima are of interglacial age and correspond in time to the soil interlayers (after Hovan *et al.*, 1989). Reprinted with permission of the American Geophysical Union.

average time span of about 3 ky during the last glacial maximum. This study showed that the intensity of the East Asian summer monsoon is prone to rapid short-term variations that may be controlled by other factors beyond the orbital forcing of the Milankovich cycles.

4.3.4   Eolian records from the North Pacific Ocean

The pelagic sediments of the North Pacific Ocean provide some of the most complete records of eolian sedimentation in East Asia over millennial time-scales because in the deep open ocean few processes are able to affect clastic sedimentation except eolian transport. One of the best studied records comes from Core V21-146, recovered on the Shatsky Rise east of Japan, which is believed to span ~530 ky based on oxygen isotope stratigraphy (Hovan *et al.*, 1989). As noted in Chapter 3, Central Asia is the principal source of eolian dust to the North Pacific and so the varying eolian flux and eolian grain size at V21-146 can be used as a proxy for continental aridity and wind strength respectively. Figure 4.14 shows how the eolian flux at V21-146 varies strongly through time and correlates with the loess–paleosol alternations on the Loess Plateau, as well as with the Marine Isotope Stages of the Late Pleistocene. This indicates that the aridity of Central Asia is controlled by the same processes that govern the Northern Hemispheric Glaciation.

Spectral analysis of the Shatsky Rise record by Hovan *et al.* (1991) showed that the eolian flux was coherent over all the Milankovich frequencies of 100, 41 and 21 ky. In contrast, the grain size of the eolian grains varies at 33 and 100 ky frequencies, suggesting that the winter monsoon strength is decoupled from the aridification process. At the 100 ky frequency, Hovan *et al.* (1991) showed that the coarser grains (and faster winds) were associated with interglacial conditions. This was a surprising result because at shorter timescales the opposite is true, although the same pattern was recognized elsewhere in the North Pacific by Janecek and Rea (1985). Hovan *et al.* (1991) suggested that these patterns might be explained if the zone of strongest winter monsoon winds were displaced south during glacial periods, so that a record at Shatsky Rise does not necessarily always sample the fastest wind strengths.

4.4   Monsoon records from cave deposits

Cave deposits in the form of stalagmites, known more generally as speleothems, have become a major tool in reconstructing the evolution of monsoon strength at high resolution. Speleothems make ideal paleoclimate indicators because they often form continuously over thousands of years and crystallize, recording the oxygen isotope character of the waters from which they formed. The oxygen isotope character reflects that of the rainwater together with equilibrium factors during calcite precipitation (e.g., flow path, CO_2 partial pressure, residence time, concentration of solutes and degassing history). Moreover, inorganic calcite can be accurately dated by mass spectrometry using ^{230}Th methods (Edwards *et al.*, 1987), which, when combined with the counting of annual banding, results in a high-resolution, well constrained climate record.

4.4.1   *Hulu Cave in Eastern China*

The speleothem deposits in Hulu Cave, located near Nanjing in central eastern China have become a classic example of how these rocks can be used to reconstruct evolving monsoon conditions (Wang *et al.*, 2001a). Most of the precipitation (around 80%) in this region now falls during the summer monsoon, when water δ^{18}O values are lower (-9 to $-13‰$) than in the winter time (-3 to $+2‰$). Thus changes in monsoon climate that shift the balance of winter and summer monsoon strength can drive large changes in the δ^{18}O values preserved in the speleothems. In contrast, the effect of changing average temperatures on water δ^{18}O values

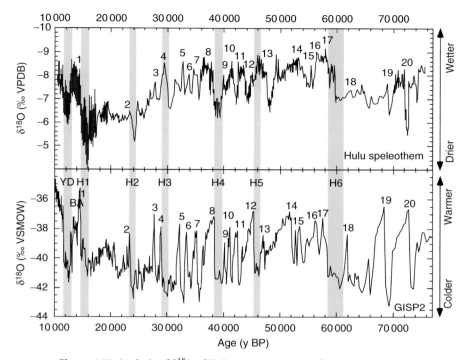

Figure 4.15 Analysis of $\delta^{18}O$ of Hulu Cave stalagmites from Wang *et al.* (2001a) and Greenland Ice Core. Numbers indicate GISs and correlated events at Hulu Cave. The Younger Dryas (YD) and Heinrich events are depicted with vertical bars. The average number of years per $\delta^{18}O$ analysis is 130–140. The $\delta^{18}O$ scales are reversed for Hulu (increasing down) as compared with Greenland (increasing up). BA = Bølling-Allerød. VSMOW = Vienna standard mean ocean water. VPDB = Vienna Pee Dee Belemnite.

appears to be quite small. Figure 4.15 shows the results of Wang *et al.* (2001a) plotted against the $\delta^{18}O$ values of the Greenland GISP2 ice core for comparison.

This study noted that the broadscale evolution of the cave $\delta^{18}O$ values followed the shape of changing summer insolation in this region, implying that the strength of the summer monsoon, which is the main control on $\delta^{18}O$ values, is largely driven by insolation changing the temperature differences between the continent and the Pacific Ocean. Because the amplitude of the $\delta^{18}O$ value change is around 5‰ this implies a change in summer precipitation by a factor of around three between maximum and minimum periods. Recent work by Cheng *et al.* (2006) on older Hulu cave records demonstrates a similar relationship with insolation during the last deglaciation, by showing a weak monsoon interval between 135.5 ± 1.0 and 129.0 ± 1.0 ka, prior to an abrupt increase in monsoon intensity.

In detail the smaller scale features from the cave record show close correlation with climatic events revealed by analysis of the Greenland GISP2 ice core

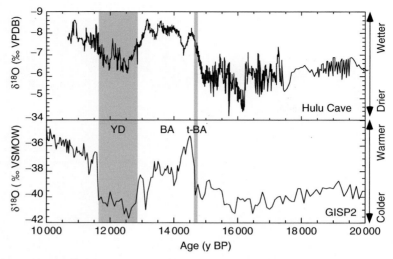

Figure 4.16 $\delta^{18}O$ of Hulu Cave stalagmites from Wang *et al.* (2001a) and Greenland GISP ice cores versus time. Gray bands indicate the timing and duration of the Younger Dryas and the transition into the Bølling-Allerød (t-BA); the BA is the interval between the gray bands. The chronology of the speleothem is fixed by annual banding. The average number of years per $\delta^{18}O$ analysis is 60 for PD, 9 for YT and 7 for H82.

(Dansgaard *et al.*, 1993). Warming events seen in the $\delta^{18}O$ values of the ice core correlate with shifts to more negative $\delta^{18}O$ values in the speleothem records, indicating wetter summer monsoons. Wang *et al.* (2001a) concluded that the climatic shifts in Greenland were synchronous within error of the age control to those seen in eastern China. The nature of the climatic shifts are not, however, identical in each area. While the change in ice core $\delta^{18}O$ values is very rapid at ~14.6 ka, during the warming into the Bølling-Allerød, this transition is more gradual in the Hulu speleothem (Figure 4.16), as it is at the end of the Younger Dryas. Conversely the start and end of the Younger Dryas cold period (13–11 ka) is very rapid in both Greenland and Hulu Cave.

Zhao *et al.* (2003) conducted a more detailed study of the Hulu Cave (which they call Tangshan Cave) speleothems since the Last Glacial Maximum. This study confirmed that the abrupt deglacial climatic oscillations from 16 800 to 10 500 y BP are semi-synchronous with those found in Greenland ice core records (Figure 4.17). Relatively rapid shifts in speleothem oxygen isotope ratios demonstrate that the intensity of the East Asian monsoon switched in parallel with the abrupt climatic transitions separating the Bølling-Allerød, Younger Dryas, and pre-Boreal climatic events. However, in all cases the dated isotopic transitions in the speleothems appear to have lasted longer, suggesting a dampened response of the monsoon to the rapid forcing that drove climate change in the North Atlantic.

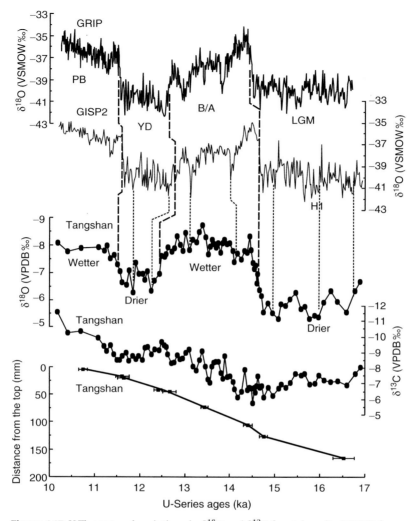

Figure 4.17 U-Th ages and variations in $\delta^{18}O$ and $\delta^{13}C$ for stalagmite 996182 from Tangshan (Hulu) Cave measured by Zhao *et al.* (2003). Ages for C-O data are calculated based on the best-fit line through the U-series age points.Shown for comparison are ice core $\delta^{18}O$ records from Greenland [GRIP (Johnsen, 1992) and GISP2 (University of Washington)]. *y* axes for the Tangshan $\delta^{18}O$ and $\delta^{13}C$ records are reversed to show the high degree of correlation with the Greenland ice cores on millennial–centennial timescales. Thick dashed vertical lines separate the LGM, BA Interstade, YD Stade and Preboreal (PB). Dotted vertical lines connect coeval century-scale oscillations, such as the H1 event. Near-constant isotope ratios of calcite samples collected along three growth layers at 17.5, 67.5, and 137.5 mm from the top of the speleothem confirm that the calcite was precipitated in oxygen isotopic equilibrium. This inference was further reinforced by the lack of significant correlation between $\delta^{18}O$ and $\delta^{13}C$ values. Normalization to Vienna Pee Dee Belemnite (VPDB) is through National Bureau of Standards NBS-19 and NBS-18. Reprinted with permission of Elsevier B.V.

The Hulu speleothem record is important in that it highlights a link between the East Asian Monsoon and North Atlantic climate and supports the idea that millennial-scale events identified in Greenland are hemispheric or wider in extent. However, what is not clear is how the connection between these systems is made. It is possible that massive iceberg calving (Heinrich) events that preceded the end of glacial periods resulted in such major changes in Atlantic circulation pattern that global ocean circulation was also affected (Broecker, 1994). Such a re-organization would influence the temperature of the Pacific Ocean and, in turn, the intensity of the East Asian summer monsoon. Alternatively, Zhao *et al.* (2003) proposed a dominant role for atmospheric teleconnections in the rapid propagation of deglacial climatic signals on a hemispheric scale. These authors supported a hypothesis by Porter and An (1995) that rapid changes in atmospheric circulation over the North Atlantic had a cascading effect on atmospheric patterns throughout the northern hemisphere, resulting in a rapid cooling of atmospheric temperatures over central Asia that increases the power of the winter monsoon, while reducing that of the summer system.

4.4.2   *Dongge Cave in SW China*

Further understanding of the East Asian monsoon comes from study of speleothems taken from the Dongge Cave in SW China (Figure 4.3), which, like the Hulu region, experiences a cool, dry winter and a wet summer monsoon. Unlike Hulu, however, the Dongge Cave is also influenced by the South Asian monsoon as well as the East Asian system. As at Hulu, age control was achieved by ^{230}Th methods and banding, resulting in an uncertainty of only ~50 y in the climate record. Wang *et al.* (2005) reconstructed a clear, long-term fall in speleothem $\delta^{18}O$ values from ~8 ka to the present, interpreted to reflect a weakening contribution from isotopically negative summer rains (Figure 4.18). As a result, the Dongge Cave record shows reducing East Asian summer monsoon strength since 8 ka. The reduced long-term trend in summer monsoon strength parallels the reduced insolation at this latitude that appears to have an immediate feedback into the temperature differential between land and ocean (Wang *et al.*, 2005).

In detail, the Dongge record shows a number of smaller amplitude, rapid events. Eight weak monsoon events, each lasting around one to five centuries and centered at 0.5, 1.6, 2.7, 4.4, 5.5, 6.3, 7.2 and 8.3 ka were identified by Wang *et al.* (2005). Each of these had an average temporal spacing of ~1.2 ky, and in many cases appear to correlate with ice-rafting Bond Events from the North Atlantic (Bond *et al.*, 1997, 2001). The events at 8.3 and 4.4 ka are seen to be the most intense and longest in duration. It is noteworthy that 8.2 ka is known as a time of major climate change in the Greenland ice cores, with a huge isotopic spike seen in the GISP2 record (Figure 4.18). This 8.2 ka event has been

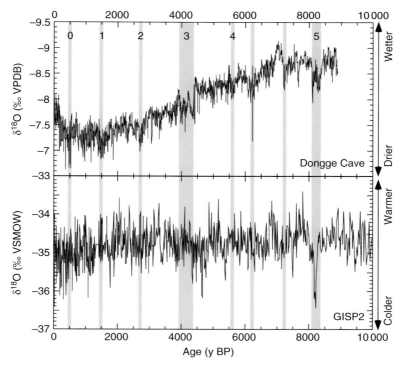

Figure 4.18 $\delta^{18}O$ of Dongge Cave stalagmites from Wang *et al.* (2005) and Greenland GISP2 ice cores plotted versus time. Gray vertical bars indicate events in the evolution of the Holocene East Asian monsoon. Numbers 0 to 5 indicate the numbers of Bond Events from the North Atlantic (Bond *et al.*, 1992, 2001).

interpreted as the result of rapid draining of a North American lake system that caused major short-term cooling as a lid of buoyant, fresh water covered large tracks of the North Atlantic, causing the thermohaline pump of the Gulf Stream temporarily to shut down (Barber *et al.*, 1999).

Wang *et al.* (2005) analyzed the variability of the Dongge speleothem $\delta^{18}O$ values and identified significant centennial periodicities of approximately 558, 206 and 159 years. These values are close to significant periods of the atmospheric $\delta^{13}C$ record (512, 206 and 148 years) that is believed to respond to changes in solar activity (Stuiver and Braziunas, 1993). As a result, the Dongge Cave record is interpreted as showing coupling between summer monsoon strength and insolation at centennial, as well as millennial, timescales.

4.4.3   Timta Cave in the Western Himalayas

An important additional control on the recent evolution of monsoon precipitation now comes from a speleothem record from the western Himalayas

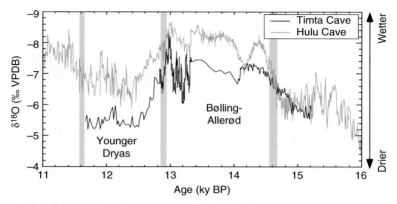

Figure 4.19 Comparison of stalagmite $\delta^{18}O$ records from Timta (black) and Hulu (gray) Caves, data from Sinha *et al.* (2005) and Wang *et al.* (2001a) respectively. Note that the caves are on independent chronologies.

at Timta Cave in western India (Figure 4.3). As described in Chapter 2, there are differences in the processes that control the intensity of the South and East Asian monsoons and it cannot be assumed that the two would evolve in identical fashions over all timescales. Differences in the evolution on centennial and millennial timescales can be determined by comparison of speleothem records from different regions. Like Hulu and Dongge, the Timta Cave is dominated by water supplied by the summer monsoon, so that isotopic variability in its record can be interpreted in a similar fashion, as reflecting the relative strength of the summer monsoon. The absolute $\delta^{18}O$ values would not be the same in Hulu, Dongge and Timta because of the different temperatures and transport histories of the moisture. Oxygen isotopes can be fractionated by the loss of moisture between the source and any given cave.

Like its Chinese equivalents, the Timta Cave record shows large isotopic variations between 16 and 11 ka (Figure 4.19). Sinha *et al.* (2005) concluded that precipitation was similar to today during the Bølling-Allerød (14.6–13.0 ka) but was significantly less during the Younger Dryas and glacial times. As at the Hulu Cave, the Timta Cave record shows a gradual wettening of the climate during the rapid onset of the Bølling-Allerød warm period, and within that period centennial-scale periodicity of drying and wettening trends are recognized. Most dramatic are the high amplitude fluctuations in $\delta^{18}O$ values seen during the latter part of the Bølling-Allerød, which are also present but reduced in East Asia. However, the fall in $\delta^{18}O$ values at the onset of the Younger Dryas is much sharper and of higher magnitude than seen in China, indicating that the summer monsoon was proportionately drier in South Asia at that time. Spectral analysis of the tuned record reveals several periods significant at the 95% (124 y, 60 y, 16 y, 13 y) and

99% (22 y) confidence limits. These short-term fluctuations have also been observed in the Late Holocene from tree ring records from the region (Stuiver and Braziunas, 1993). This correspondence indicates that similar multi-decadal forcing processes were operating during the end of the Last Glacial Maximum and the Holocene (Gupta et al., 2005), even though the boundary conditions of these two time intervals were different. Variations in insolation on these short timescales are the most likely controls on monsoon strength in South Asia since the Last Glacial Maximum.

4.4.4   *Cave records around the Arabian Sea*

Three principal locations have been used to study varying monsoon precipitation in the Arabian Sea region, the Hoti Cave in northern Oman, the Qunf Cave in southern Oman and the Moomi Cave on Socotra Island offshore of Somalia (Figure 4.3). In a pioneering study, Burns et al. (1998) examined two stalagmites and a "flowstone" from Hoti Cave and dated them using the U-Th method (Edwards et al., 1987). The speleothems had quite different ages of 125–120 ka, 117–112 ka and 9.7–6.2 ka, representing different phases of rapid speleothem growth, but all restricted to interglacial periods. The formation of a flowstone, together with its isotope character, indicated very wet summer monsoon conditions at 125–120 ka, an interglacial period. After the end of flowstone sedimentation-stalagmites formed, with $\delta^{18}O$ values rising from between -9 and $-12‰$ to between -6 and $-9‰$ after 117 ka, pointing to reduced, if still high, rates of summer monsoon precipitation. Curiously, no material was found covering the time period 112 to 9.7 ka, possibly indicating conditions too dry to form stalagmites and consistent with the idea of a weak Asian summer monsoon during glacial periods. From 9.7–6.2 ka large stalagmites again grew quickly and show $\delta^{18}O$ values of -6 to $-4‰$. Although these data clearly show wetter conditions in the Early Holocene in Arabia it is noteworthy that both growth rates and $\delta^{18}O$ values show that this time was not as wet as at 117–125 ka. Since 6.2 ka, $\delta^{18}O$ values are similar to those now forming in Oman, consistent with a generally arid climate during the Late Holocene after the Early Holocene maximum.

The evidence for a wet Early Holocene climate in Arabia correlates with the data from Hulu and Dongge Caves, as well as pollen data from east Africa (van Campo et al., 1982). Oceanographic proxies are partly in accord with strong summer monsoon winds during interglacial periods, but typically provide no information about continental precipitation. It is important to remember that marine indicators of enhanced biogenic production and increased upwelling, which have been interpreted to reflect enhanced monsoon strength, are not necessarily coupled to the position of the intertropical convergence zone (ITCZ), which is also important in controlling precipitation. Monsoon intensification, by

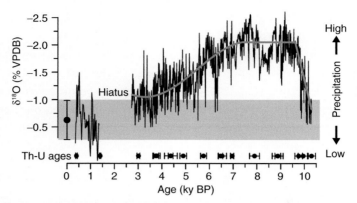

Figure 4.20 Q5 stalagmite $\delta^{18}O$ records from the Qunf Cave in southern Oman from Fleitmann *et al.* (2003). Black dots with horizontal error bars are TIMS and MC-ICPMS Th-U ages. The black dot with vertical error bars and the gray shaded area show the $\delta^{18}O$ range of modern stalagmites (101 stable-isotope measurements during the past 50 years). The heavy gray line shows the long-term trend defined for the early Holocene and a sinusoidal in the late part. Reproduced with permission of American Association for the Advancement of Science.

itself, although leading to increased upwelling along the Oman margin, might not bring more moisture to Oman if the intertropical convergence zone does not shift northward at the same time.

The Qunf Cave in southern Oman provides an opportunity to study the monsoon in southern Arabia and crucially now sits at the present northern limit of the summer migration of the ITCZ and the associated Indian Ocean monsoon rainfall belt. More than 90% of total annual precipitation (400–500 mm) here falls during the monsoon months (July to September). Fleitmann *et al.* (2003) analyzed a series of bands from a stalagmite in the Qunf Cave spanning the last 10 ky. Figure 4.20 shows the monsoon evolution pattern with a coherent long-term pattern of decreasing $\delta^{18}O$ values after 10 ka, indicating increasing summer monsoon strength after the Younger Dryas cold, dry phase. As noted in the Timta Cave, all stages of the monsoon evolution show strong decadal to centennial variations, interpreted as being driven by changes in the solar activity. The $\delta^{18}O$ values remain low and indicative of a maximum in the summer monsoon intensity from 9.6 to ~7.0 ka. After this time, $\delta^{18}O$ values increased as the summer monsoon rains weakened until 2.7 ka. No stalagmite formation occurred from 2.7 to 1.4 ka, possibly because the climate had desiccated to the point that cave water flow ceased. The $\delta^{18}O$ values in the 1.4–0.4 ka period are similar to modern values and indicate a relatively stable, arid environment in southern Oman since that time. Although monsoon winds are strong in this

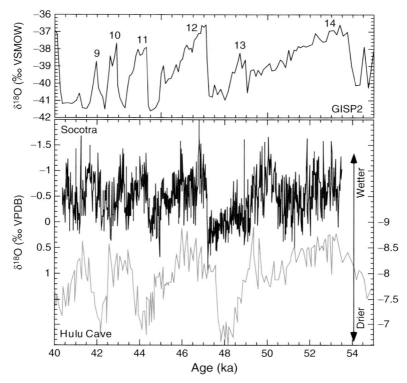

Figure 4.21 $\delta^{18}O$ records for the speleothem on Socotra Island from Burns *et al.* (2003) showing the clear parallels between monsoon precipitation records on Socotra and at Hulu Cave in eastern China. Both records appear to correlate closely with variations from the GISP2 ice core, indicating a dominant insolation control on monsoon strength across Asia.

region, the intertropical convergence zone lies south of the region and prevents significant precipitation from driving much further speleothem formation.

A third detailed record of Late Pleistocene climate evolution has been derived by analysis of speleothems from Socotra Island. Because rainfall on the island is related to the intensity of tropical convective activity, it is likely representative of the tropical hydrological cycle in the Indian Ocean, and perhaps to variations in monsoon rainfall over a much larger area. Although again there is considerable short-timescale variation, millennial-scale variation in the isotopic record strongly resembles the pattern of isotopic variation in the Greenland ice cores (Figure 4.21) and the Hulu Cave. Burns *et al.* (2003) interpreted this parallel evolution to reflect teleconnections between the North Atlantic glacial system and the Arabian Sea. Clearly, insolation forcing of the Asian monsoon through orbital variations has been occurring at a variety of timescales throughout the Holocene–Late Pleistocene, influencing summer monsoon precipitation across Asia.

4.5   Monsoon variability recorded in ice caps

Long-lived ice sheets have developed within the influence of the Asian monsoon and as in Greenland and Antarctica these can be exploited for paleoclimatic reconstructions. Because of the low latitude of much of the monsoon region, ice sheets are not common but are found in places around the Tibetan Plateau and the Himalayas. The Guliya ice cap from the western Kunlun Shan in western Tibet (Figure 4.3) receives most of its precipitation during the summer monsoon and is a good location to examine long-term precipitation variability. The ice cap has been dated by cosmogenic isotope (^{36}Cl) methods back to 500 ka at its base and can potentially provide a high resolution, long-term history spanning that time interval (Thompson et al., 1997). However, because of deformation in the ice due to its progressive motion those parts of the ice stratigraphy deposited before ~130 ka are not useful for paleoclimatic work. Nonetheless, the ice was analyzed for a series of proxies to reconstruct monsoon strength since 130 ka, including eolian dust accumulation rates, as well as Cl^-, NO^{3-} and SO_4^{2-} concentrations. Both Cl^- and SO_4^{2-} are typically believed to be transported as dust from salt flats and lake beds and could be used to trace winter monsoon intensity. In contrast NO^{3-} likely originates from soils and vegetation and would be more linked to summer monsoon precipitation. The $\delta^{18}O$ values measured in the ice reflect the $\delta^{18}O$ values of the precipitation, and to a large extent the temperature of the air over Tibet during the geologic past.

As with cave deposits the $\delta^{18}O$ evolution of the Guliya ice shows a first order correlation with the Greenland and Antarctica ice core records (Thompson et al., 1997), reflecting the long-term changes in global climate during the last glacial period (Figure 4.22). The $\delta^{18}O$ values in the Guliya ice decreased during glacial periods and were at their lowest levels during the Last Glacial Maximum (Figure 4.22). The similarity of the $\delta^{18}O$ evolution to that of CH_4 records from polar ice cores was interpreted by Thompson et al. (1997) to indicate that global CH_4 levels and the tropical hydrological cycle are linked, though it is not yet clear whether one is controlling the other, or if they are both forced by some other system, likely insolation. Clearly, the monsoonal climate of western Tibet must have been cool and quite variable during the Last Glacial Maximum.

In addition to the long-term evolution, the Guliya ice cores demonstrate that the Late Glacial Maximum record contains numerous 200 y scale oscillations in $\delta^{18}O$ values, as well as in dust, NH_4^+, and NO_3^- levels, indicating forcing by other processes beyond the longer term orbital Milankovich cycles. From 15 to 33 ka, over 100 such variations in $\delta^{18}O$ values, spanning 2 to 21‰ are noted and these are likely linked to sunspot activity, which is known to vary on such timescales

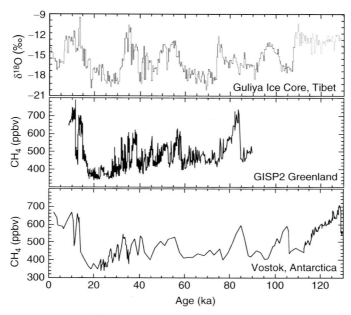

Figure 4.22 The $\delta^{18}O$ records for the Guliya ice cap of western Tibet are shown with time for the past 130 ky. Data from Thompson *et al.* (1997). The record is compromised by ice deformation below 110 ka. The Guliya $\delta^{18}O$ record over the past 110 ky is matched to the CH_4 records from GISP2 and the Vostok icecore of Antarctica (Petit *et al.*, 1999, 2001) to assess possible linkages.

and could affect South and East Asia, Antarctica and Greenland. However, the highest isotopic variations are too great to be explained only as changes in temperature, and probably also require changes in the location of the high pressure atmospheric cell that sits over the Tibetan Plateau, as a result of changing surface temperatures. In practice, periods of higher insolation may have caused melting and reduction of Tibetan ice and snow fields, increasing plateau summer heating. Exposure of the soil also allows more dust to be entrained into the winds, thus increasing the measured NH_4^+, and NO_3^- levels.

4.6   Monsoon variability recorded in lacustrine sediments

Long-lived lakes form natural repositories of information on continental climatic evolution because they are basins of relatively continuous sedimentation. Because their water columns are often strongly stratified the bottoms are usually anoxic and thus not bioturbated. Consequently, very high resolution fine-grained sedimentary sequences may accumulate. Moreover, precipitation and run-off are major controls on the nature of lake sedimentation, controlling

Zabuye Lake, western Tibet

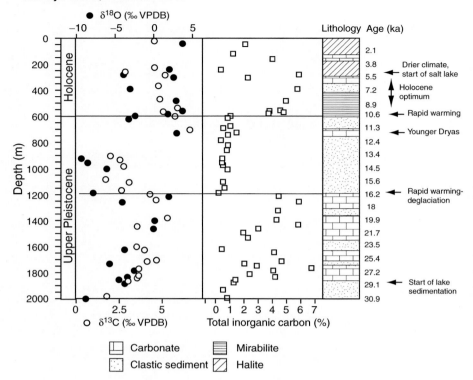

Figure 4.23 Comparison of carbon and oxygen isotope values with total inorganic carbon concentrations of core ZK2 from the Zabuye Lake for the last ~30 ky and interpretation of paleoenvironmental, climatic and hydrological features over the last glacial–postglacial and Pleistocene–Holocene transitions. The entire core ZK2 is separated into three major sections based on the lithology. Both carbon and oxygen isotope values reported in ‰ vs. PDB. Redrawn from Wang *et al.* (2002).

both lake levels and the nature of the clastic flux reaching the basin. Lakes in the region of Tibet can be used to track the development of Holocene–Late Pleistocene climate in that region. Lake Zabuye is a salt lake located at 4421 m elevation within the Transhimalayan ranges of SW Tibet (Figure 4.3). The lake has been cored to a depth of 20 m, extending back into the Upper Pleistocene on the basis of ^{14}C and ^{210}Pb ages (Wang *et al.*, 2002; Figure 4.23). Although the climate is cool, humidity and rainfall are both very low, which leads to a dominantly evaporative state, accentuated by the fact that the lake is recharged by flux from other evaporating lakes.

Wang *et al.* (2002) examined the lithology of the core and analyzed material for total organic and inorganic carbon, as well as oxygen and carbon isotopes, in an attempt to quantify the evolving climate around Lake Zabuye. As for the ice cores the $\delta^{18}O$ values measured from bulk carbonate samples are interpreted to reflect the temperature of the air over western Tibet and to a lesser extent the amount of summer monsoon precipitation. In addition, the $\delta^{18}O$ values of carbonate precipitated in the lake also reflect the degree of evaporation, which will tend to make the $\delta^{18}O$ values more positive.

The stratigraphy of the core is easily subdivided into three major sections: (1) a basal carbonate-clastic mixture, (2) a middle clastic-dominated sequence and (3) an upper sequence of carbonate and evaporite minerals (mirabilite and halite). The age control allows us to see that the transition from carbonate into more clastic sediment corresponds to ~16.2 ka, a period of rapid global warming. The faster sediment flux and lower $\delta^{18}O$ values suggest a stronger summer monsoon and more run-off into the lake, temporarily disrupted by a period of drying and weaker summer monsoon during the Younger Dryas. Water flow into the lake was probably very high at 16.2–10.6 ka, owing to an intensifying summer monsoon and because of the melting of glaciers. It is interesting to note that carbonate $\delta^{18}O$ values increased again after 10.6 ka even though this represents the start of the Early Holocene climatic optimum, which in other areas represents a period of strong summer monsoon. This may be the case here too, but as temperatures increased after the cold Younger Dryas evaporation also increased, resulting in a net shift to more positive $\delta^{18}O$ values in the lake carbonate.

Carbonate contents rose sharply during the Early Holocene, along with $\delta^{13}C$ values, thus suggesting a surge in biogenic production under the influence of a strong summer monsoon, which slackened again after 5.5 ka. This is the time at which evaporites began to form and together these data point to lower degrees of summer monsoon precipitation. The region became severely desiccated around 3.8 ka when the modern saline lake was established. Wang *et al.* (2002) argued that $\delta^{13}C$ values were controlled partly by water balance in the lake, because this affects the relative amount of carbon derived for organic versus atmospheric sources. The fall in $\delta^{13}C$ values after 16.2 ka is thus interpreted to indicate lower residence times as water flow increased, climbing again as the region dried during the Holocene. Like $\delta^{18}O$, $\delta^{13}C$ values are quite variable over short time intervals in the Early Holocene owing to a rapidly varying climate of warmer and cooler phases at that time.

It can be seen that the Zabuye Lake record shows large changes in monsoonal climate over the western Tibetan Plateau since the Last Glacial Maximum, and that these changes correlate quite closely with global climate change. There was an initial strengthening of the summer monsoon after a dry glacial climate, then subsequent weakening in a warm climate after the Younger Dryas. The lake is too much in the rain shadow of the Himalayas to exhibit a summer monsoon comparable with those seen in the plains of the Indian sub-continent.

A similar record can also be derived from coring of Lake Luanhaizi in NE Tibet–China close to the Xilian Shan (Mischke *et al.*, 2005). Unlike Zabuye Lake, this region lies in the influence of the East Asian monsoon and represents an important additional control to the Late Pleistocene–Holocene climate, beyond what is known from the nearby Loess Plateau. Lake Luanhaizi lies close to Qinghai Lake, but differs in being much smaller and more Alpine in character. This lake lies at 3200 m elevation and appears to have a record dating back to at least 45 ka, based on combined ^{14}C AMS and magnetic stratigraphy studies on two cores penetrating ∼14 m below the modern lake floor. The cores were analyzed for a suite of major elements (Mischke *et al.*, 2005), as well as for the ostracod and aquatic plant record (Herzschuh *et al.*, 2004, 2005), which is strongly controlled by the level and salinity of the lake, and thus in turn by summer monsoon intensity (Figure 4.24).

Mischke *et al.* (2005) used a combination of these proxies to divide the cored section into nine stages (Z1 to Z9). They argued that for three identified stages, Z2, Z6a (corresponding to Marine Isotope Stage 2) and Z9 (Holocene), the lake was deeper than it is at present. In contrast, the same data indicate a shallower or even a playa lake during stages Z4, Z6b and Z8 (Younger Dryas). At first order the lake is seen to have been deeper and thus the summer monsoon stronger during interglacial periods, with drier conditions during glacial times. The sandy sediments of stage Z5 are assigned to the Last Glacial Maximum, so that stage Z6 would represent the initial deglaciation, which is marked by rapid, pulsed influxes of meltwater from glaciers. It is noteworthy that no glaciers covered the lake even during the Last Glacial Maximum. The lake seems to have dried up entirely during deposition of stage Z7, only to fill again during the Bølling-Allerød, then dry and become more saline during the Younger Dryas. The lake appears to have been deeper than today during the Early Holocene, consistent with the idea that that time was one of a stronger summer monsoon precipitation. The results are consistent with those reported from Lakes Aksayqin (Li *et al.*, 1989), Dachaidan (Huang *et al.*, 1980), and Lop Nur (Yan *et al.*, 1983) in NW Tibet and the Tarim Basin, as well as Tsokar Lake in Ladakh, India (Bhattacharya, 1989) in indicating stronger summer monsoon rains during interglacial periods.

Lake records from the Indian foreland are not as well documented as those in China and Tibet, but are still valuable in the reconstruction of monsoon evolution because the summer rains are so strong in this area. Sanai Lake is located in the central flood plain of the Ganges River and has received special study because of the acquisition of cores that allow its evolution to be documented. Sharma *et al.* (2004) have documented the development of the lake

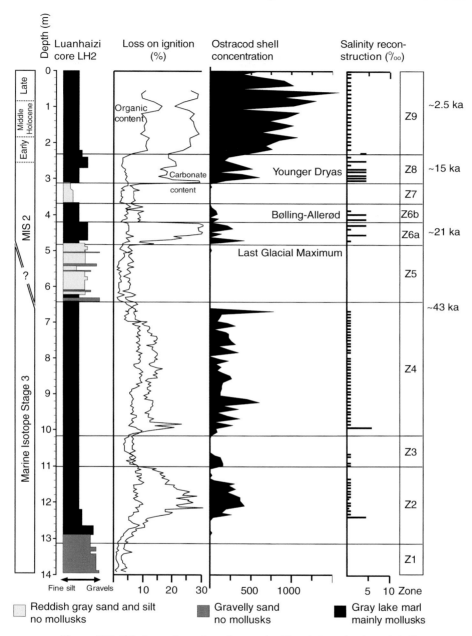

Figure 4.24 Lithology, chronology, loss on ignition data and ostracod shell concentration of core LH2 from Lake Luanhaizi measured by Mischke *et al.* (2005). Reconstructed salinity ranges encompass the past salinity level based on salinity tolerance ranges of taxa present at a sampled level. Salinity ranges are not given for ostracod count sums <40. Reprinted with permission of Elsevier B.V.

since 15 ka. Stable isotope, elemental geochemistry and pollen analysis were used to identify periods of greater aridity, with these conditions being seen at 15–13 ka, 11.5–10 ka and 5–2 ka. The 11.5–10 ka period corresponds with the Younger Dryas and is marked by a distinct decline in all plant taxa, including trees, shrubs, aquatic taxa, herbs and ferns, except for grasses and *Botryococcus* sp. that exhibit an increasing trend. During the Early Holocene a deep, permanent lake was established, testifying to a strong, wet summer monsoon from 10–5.8 ka, after the end of the Younger Dryas. Plant species diversified significantly during that time, and Sharma *et al.* (2004) report a high prevalence of aquatic elements, such as *Potamogeton* sp., and *Typha* sp., together with a decline in sedges and ferns. This pattern further suggests that Sanai Lake expanded considerably after 10 ka under the influence of a strong summer monsoon. Nonetheless, strong variation in oxygen and carbon isotopes again indicate monsoon strength changing quickly on nonMilankovich timescales during that period. The dry climate prior to 2000 y BP began to wetten again after 1700 y BP.

4.7   Salinity records in marine sediments

In East and Southeast Asia the summer monsoon is characterized by heavy precipitation and thus run-off from the continents to the oceans. The large-scale discharge of fresh water from rivers into the ocean must necessarily change the salinity locally (Duplessy, 1982), and this in turn can be used as a proxy for monsoon intensity if salinity can be calculated. Salinity is one of the controls on the oxygen isotope composition of foraminifer tests, but this cannot be uniquely determined without additional constraints on the paleo-sea-surface temperatures. Sea-surface temperature estimates can be derived using the biomarker U^K_{37} index for alkenones, which provides an estimate of sea-surface temperature annual mean at 0–30 m water depth (Kirch, 1997; Pelejero and Grimalt, 1997). Based on five replicate extractions of alkenones from sediment samples, the error of these temperature estimates is only about 0.15 °C. Paleosalinities of surface water can then be estimated by subtracting the past variations in sea-surface temperatures and global ice volume from the planktonic $\delta^{18}O$ signal of the foraminifer *Globigerina ruber*.

Kudrass *et al.* (2001) used sediment from core site SO93-126KL offshore of the Ganges-Brahmaputra delta to reconstruct the varying run-off from the delta over the last 80 ky (Figure 4.25). In this case the variation in sea-surface temperature has been very modest and so the $\delta^{18}O$ values vary closely with salinities. The salinity record closely follows the GISP2 Greenland ice record, indicating a close coupling between monsoon intensity in the eastern Himalayas and glacial intensity in the North Atlantic, with each Dansgaard–Oeschger event apparently

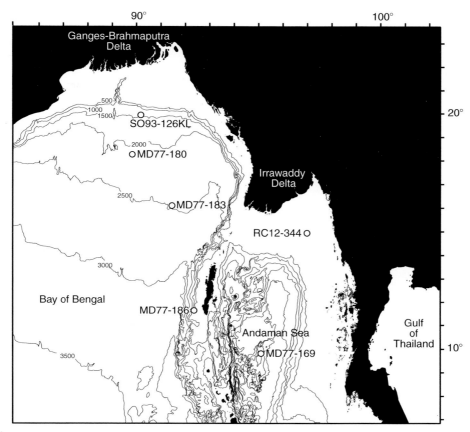

Figure 4.25 Bathymetric map of the Bay of Bengal and the Andaman Sea showing the location of coring stations mentioned in the text. Data are from GEBCO compilation. Water depths in meters.

mirrored in the Bengal salinity record (Figure 4.26). Salinity was greatest during the Last Glacial Maximum, indicating a strongly reduced fluvial run-off and thus low precipitation. The rapid decreases of $\delta^{18}O$ values starting at 14.2 ka and 11.8 ka correlate with the global melt water spikes MWP1A and MWP1B and the associated rises in sea level (Fairbanks, 1989; Siddall et al., 2003; Camoin et al., 2004). The low U^k_{37} temperatures and the shifts by 1.0‰ and 1.5‰ in foraminifer $\delta^{18}O$ values indicate an addition of cold, oxygen isotope-light melt water from Himalayan glaciers in the Bay of Bengal after 14.2 ka. The Younger Dryas (13–11 ka) period exhibits the expected return to drier conditions similar to the Last Glacial Maximum (20 ka). In the middle Holocene, $\delta^{18}O$ values reached a minimum owing to a maximum in summer monsoon intensity, which has subsequently declined since ~5.0 ka (Feng et al., 1999; Sirocko et al., 1996).

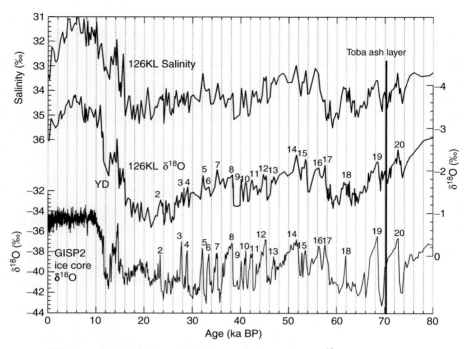

Figure 4.26 Correlation of salinity changes and marine $\delta^{18}O$ values from planktonic foraminifera *Globerigerinoides ruber* in northern Bay of Bengal with Greenland GISP2 $\delta^{18}O$ ice record (Dansgaard *et al.*, 1993). Salinity is calculated from marine $\delta^{18}O$ and U^k_{37} temperatures (from Kudrass *et al.*, 2001). Numbers indicate Greenland interstadial IS 1–20; YD is Younger Dryas. A layer of volcanic ash from Toba megaeruption, and its traces in Greenland ice core (Zielinski *et al.*, 1996), is used as an independent time marker for correlation. Reproduced with permission of the Geological Society of America.

Salinity reconstructions stretching back 40 ky have also been made in the South China Sea, (Wang *et al.*, 1999a). The U^k_{37} biomarker temperature–salinity methodology was also employed in this setting, and was applied to sediment taken from core 17940-1/2 recovered from the northern margin of the South China Sea, not far from the estuary of the Pearl River. As in the Bay of Bengal the record calculated by Wang *et al.* (1999a) shows a general correlation between periods of glaciation and major changes in the salinity of the ocean (Figure 4.27). Salinity was low during the warm Bølling-Allerød and during the subsequent Preboreal and Early Holocene periods, with higher salinity seen during the Younger Dryas. This pattern was interpreted by Wang *et al.* (1999a) to reflect variations in precipitation and run-off from southern China, with drier conditions associated with the colder Younger Dryas. The Last Glacial Maximum also appears to have been a time of greater salinity, which together with the general

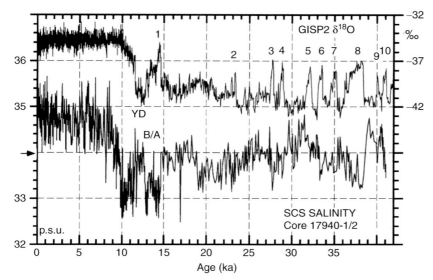

Figure 4.27 Close-up of sea surface salinity variations calculated by Wang *et al.* (1999a) at coring site 17940 on the northern margin of the South China Sea, which are generally reversed to the δ^{18}O-temperature oscillations in the GISP2 ice record (Grootes and Stuiver, 1997). Numbers are Dansgaard–Oeschger events. Reprinted with permission of the American Geophysical Union.

lack of fluvially derived sediments indicates a minimum of run-off and presumably great aridity in southern China at that time.

It is noteworthy that salinities for the South China Sea during the Last Glacial Maximum were 1.0–1.5‰ lower than today, which might imply more precipitation at that time. However, it should be noted that salinity at the Last Glacial Maximum was >1‰ higher than during the Bølling-Allerød. The lower salinity during the Last Glacial Maximum probably reflects the reduced interchange of the South China Sea with the Pacific Ocean, because lower sea levels closed or reduced the oceanic gateways that now provide interchange with more saline open ocean waters. In addition, because the sea level was lower, the mouth of the Pearl River must have been closer to the core site, thus delivering fresh waters in more concentrated form to the region of the core. It is noteworthy that the salinity record in the Bay of Bengal (Kudrass *et al.*, 2001) shows a salinity minimum due to heavy run-off ~6–8 ka, while salinities in the South China Sea were already relatively elevated at that time.

Prior to the Last Glacial Maximum there is a common inverse relationship between glacial intensity recorded by the Greenland GISP2 core and the salinity of the South China Sea. The rapid warming associated with Dansgaard–Oeschger event 8 (~38 ka) is especially dramatic in showing freshening of the surface waters as the glaciers retreated and the summer monsoon strengthened. Wang *et al.* (1999a) noted that the glacial periods were also times at which eolian sedimentation was strong, as well as salinities being high, indicating strong winter monsoon winds, but weak summer precipitation.

The evolving paleoceanography, and especially sea-surface salinities, of the nearby Sulu Sea in relation to varying monsoon strength were investigated by Dannenmann *et al.* (2003) using Core MD97-2141 recovered from 2633 m water depth in the Sulu Sea (Figure 4.3). The region lies just south of the South China Sea and is affected by heavy summer monsoon rains, with less rainfall in the winter. Direct rain and run-off from the surrounding islands results in reduced salinities in surface waters during the summer, a signal that can be preserved in marine sediments. Conkright *et al.* (1998) have shown that salinities in the Sulu Sea are ~0.5‰ less in October after the summer monsoon than in the winter. Unlike the South China Sea the Sulu Sea surface temperatures are not very variable between summer and winter (~2 °C) because the basin is not affected by the cooler currents that influence its northern neighbor (~5 °C variation).

Dannenman *et al.* (2003) used Mg/Ca ratios to constrain the changing sea-surface temperatures (1.0–1.5 °C) and then applied these figures to the evolving $\delta^{18}O$ values of the foraminifers to calculate sea-surface salinities. These workers reconstructed a record that showed strong coherent changes, especially during the period 30–60 ka (Figure 4.28), i.e., during Marine Isotope Stage 3. The amplitudes of the measured $\delta^{18}O$ fluctuations (0.4–0.7%) exceed those that can be attributed to sea level changes and this must be a result of changes in sea-surface conditions. However, it was noteworthy that Mg/Ca derived temperature estimates did not change in parallel with the $\delta^{18}O$ values, implying that it was changes in sea-surface salinity that triggered the evolving $\delta^{18}O$ values. These are at short timescales, below the shortest 21 ky period of the orbital forcing. However, the changes correlate quite well with isotopic variations seen in the Hulu Cave in central East China (Wang *et al.*, 2001a), demonstrating that precipitation changes in the Pleistocene Sulu Sea were controlled by varying strengths of the summer East Asia Monsoon.

Dannenmann *et al.* (2003) noted that, within dating uncertainties, the suborbital variability in Sulu Sea $\delta^{18}O$ values indicates that times of fresher surface conditions in the Sulu Sea coincide with similar conditions in the Western Pacific Warm Pool (Stott *et al.*, 2002). Figure 4.28 shows that the salinity and precipitation records from East Asia and the Sulu Sea are coupled to the major changes seen in GISP2 Greenland Ice core records, consistent with the hypothesis that suborbital variability in the summer Asian monsoon and Intertropical Convergence Zone was tightly coupled with climate conditions in the high latitudes.

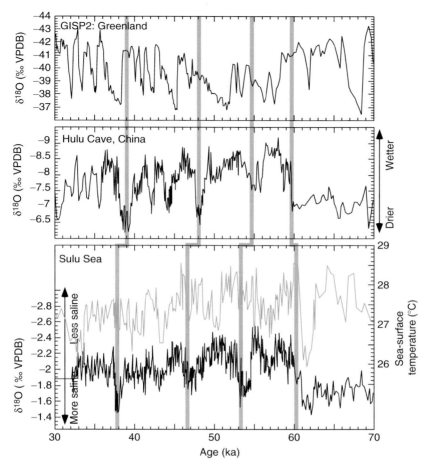

Figure 4.28 Oxygen isotope (black line) and sea-surface temperature (gray line) calculated from the Mg/Ca record for IMAGES core MD97-2141 from the Sulu Sea (data from Oppo *et al.*, 2003; Dannenmann *et al.*, 2003), compared with records from the Hulu Cave in central China (Wang *et al.*, 2001a) and the GISP2 ice core from Greenland.

4.8   Pollen records in marine sediments

The use of pollen from lacustrine sedimentary sequences as a monsoon proxy was discussed above, but here we also note that pollen can be extracted from marine sediments and used to reconstruct changing flora onshore in the adjacent landmass. Because types of vegetation are strongly controlled by the total precipitation and seasonality of the climate (especially the length of the dry season), it follows that flora should be closely linked to monsoon intensity. Furthermore, because pollen are often well preserved in sediments over geological time spans they may be used to reconstruct the nature of floral changes.

Sun and Li (1999) studied the pollen types from the same sediments recovered by Core 17940-1/2 in the South China Sea (Wang *et al.*, 1999a,b).

Three zones were recognized based on the assemblages of pollen found (Figure 4.29). The oldest Zone P1 (37.0–15.0 ka) is characterized by an abundance of conifer spores (*Tsuga, Picea* and *Abies*) that are usually associated in modern China with mountainous, cooler weather regions. In addition, the core showed many pollens indicative of temperate grassland, mainly *Artemisia*, which likely grew on the exposed northern continental shelf of China. Frequent natural fires are inferred on the basis of high, albeit variable charcoal concentrations. Sun and Li (1999) interpreted these data to indicate a cool and dry climate during 37.0–15.0 ka, which was a period of strong glaciation. As seen in the Greenland GISP2 core glacial intensity was not however constant, and this is reflected in the frequent alternations of montane conifers and *Artemisia*-dominated grassland that testify to changing cool and humid conditions with temperate and dry conditions.

A sharp change in the pollen assemblage was noted in Zone P2, corresponding to climatic warming of the Bølling-Allerød (~13 ka). A rapid rise in sea level drowned the grasslands of the South China Shelf, which were then replaced by mangrove swamps (*Rhizophora* and *Sonneratia*; Figure 4.29). The warming also caused expansion of tropical–subtropical broadleaved forest. Subsequently, climatic cooling from 13–11 ka during the Younger Dryas cold spell is recorded by expansion of both montane conifers and upper montane rain forests (mainly *Podocarpus* and *Dacrydium*). The youngest pollen Zone P3 shows a strong similarity between its pollen assemblages during the Holocene and in the surface sediment from the modern northern South China Sea, which implies that the vegetation and climate during the last 10 000 years was close to those of the present.

A parallel record has been generated from the western Indian continental margin of the Arabian Sea. Core SK128 A-31 was collected on the flanks of the Maldive–Laccadive Ridge and penetrate sediments that date back to ~200 ka. Continental pollen records from lakes in Rajasthan, India, already provided evidence for climate change since the Last Glacial Maximum (Bryson and Swain, 1981; Swain *et al.*, 1983) and the general patterns of development were confirmed by this marine core. Using pollen data from western Indian lakes, Bryson and Swain (1981) reconstructed the history of summer rainfall and concluded that the monsoon was at its peak during the Early Holocene, but reduced by about two-thirds at the present time. The dating of Core SK128 A-31 was made possible by a combination of ^{14}C ages and oxygen isotope stratigraphy (Prabhu *et al.*, 2004), allowing accurate comparison between pollen changes in the core and global climate change (Figure 4.30).

Analysis of the extracted pollen shows that salinity-tolerant vegetation (e.g., *Poaceae* and *Chenopodiaceae* or *Amaranthaceae*) is the dominant type. In addition,

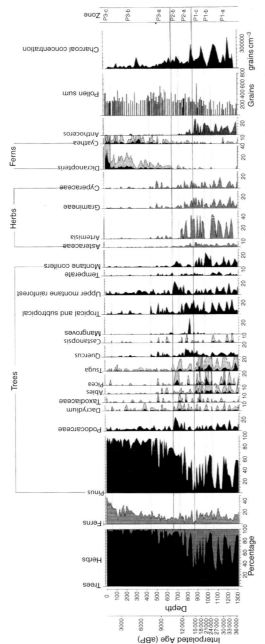

Figure 4.29 Pollen record since 37 ka for the deep water core site 17940-1/2 on the northern margin of the South China Sea (from Sun and Li, 1999). Reprinted with permission of Elsevier B.V.

137

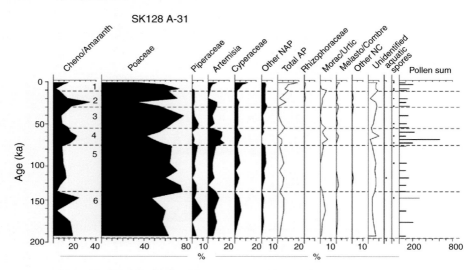

Figure 4.30 Pollen diagram for core SK-128A-31 on the western continental margin of India. Bold numbers are marine isotopic stages and dashed lines the boundaries between them (from Prabhu *et al.*, 2004). AP = arboreal pollen, NAP = nonarboreal pollen. Reprinted with permission of Elsevier B.V.

forms characteristic of evergreen forests (e.g., *Gnetum, Syzygium, Mallotus, Macaranga, Olea glandulifera*) and of deciduous forests and woodlands (e.g., *Azadirachta, Lannea* and *Combretaceae* or *Melastomataceae*) were identified. However, mangrove pollen is present only in the top few centimeters of Core SK128 A-31, and even then in very small quantities, reflecting the significant distance from the coast (Figure 4.30). *Poaceae* and *Chenopodiaceae* or *Amaranthaceae* are particularly dominant during glacial periods (Marine Isotopic Stages 2, 4 and 6), suggesting that the climate was cold and dry and that salinity-tolerant vegetation colonized large areas near the seashore because the lower sea level exposed the continental shelf, similar to that inferred by Wang *et al.* (1999a) in the South China Sea.

Artemisia is a taxon that now grows in semi-arid climates and at low elevation and is an indicator of low precipitation (Rossignol-Strick *et al.*, 1998). Its variability and prominent maxima in the Indian Ocean pollen spectra during glacial periods support the hypothesis of low summer monsoon rains during those periods. In contrast, during interglacial periods (e.g., Marine Isotopic Stages 1, 3 and 5; Figure 4.30) the arid taxa were sparse, and *Poaceae* and *Piperaceae* were abundant, so that high precipitation is inferred for southern India during those periods. It is noteworthy that Prabhu *et al.* (2004) also used organic carbon measurements to show that paleoproductivity was at a maximum during glacial times. Because upwelling offshore of western India is driven by the NE winter monsoon this is consistent with the hypothesis that the winter monsoon was stronger during glacial times.

4.9   Paleoproductivity as an indicator of monsoon strength

4.9.1   Nitrogen isotopes

Monsoon winds in many locations can cause coastal upwelling and enhanced biogenic productivity that can be preserved in the sea-floor sediments in the form of a number of proxies. The long-term evolution of upwelling-related foraminifer *Globigerina bulloides* was discussed for the Oman margin in Chapter 3. Here we look at other proxies that have been studied at high resolution to assess the millennial-scale and shorter evolution of upwelling patterns in relation to monsoon strength. Nitrogen isotopes have become an effective tool for examining oceanographic processes because denitrification occurs under suboxic conditions when bacteria process nitrate and convert it primarily to N_2 gas. This process is the primary loss mechanism for combined nitrogen from the biosphere, and thus has an important role in the nitrogen cycle and those biogeochemical cycles linked to it. In the ocean, denitrification occurs in organic-rich continental margin sediments and in intermediate waters within oxygen-minimum zones, such as those that exist in the western Arabian Sea and Oman margin. Denitrification strongly fractionates nitrogen isotopes, leaving the remaining nitrate enriched in ^{15}N (Cline and Kaplan, 1975). A paleoceanographic record for denitrification intensity may be created and then preserved when ^{15}N-enriched nitrate is transported to surface waters and consumed by plankton. Subsequently this is transported to the sea floor with organic matter and preserved in the sediments. In oxygen-minimum zones the preservation potential of this organic matter is high.

Altabet *et al.* (1995, 1999) used material from cores offshore of southern Oman (Core RC24-14; Figure 3.3) to demonstrate a strong coupling between variations in denitrification intensity and climate change at orbital periodicities (Figure 4.31). These data show a close linkage between higher $\delta^{15}N$ values (i.e., more denitrification) and interglacial periods. Denitrification requires suboxic conditions, which depends on the intensity and extent of the subsurface oxygen minimum zone. This means that $\delta^{15}N$ values largely reflect local productivity and a downward flux of organic carbon to the sea floor, as well as the speed of convective overturn of Arabian Sea water. In practice, this implies that $\delta^{15}N$ values reflect paleoproductivity and thus the intensity of summer monsoon-induced upwelling. The nitrogen isotope record from the Oman margin is in accord with the hypothesis of monsoon strength being controlled by orbitally controlled insolation. The record indicates stronger summer monsoon winds correlating with interglacial periods.

More recently, Altabet *et al.* (2002) were able to show that the nitrogen isotopes, as well as nitrogen content and chlorin abundances offshore of Oman, are linked to glacial cycles at the centennial level. In particular, chlorin levels,

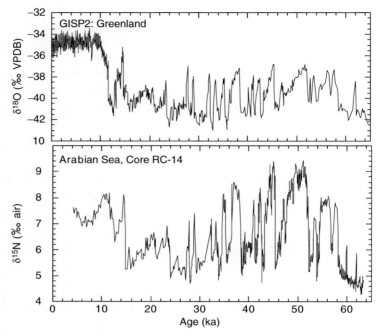

Figure 4.31 Record of δ^{15}N changes from core RC27-14 offshore of Oman, showing the close links between glacial–interglacial cycles and the denitrification process in the Arabian Sea (data from Altabet *et al.*, 1995, 1999).

which are closely linked to productivity in near-surface waters, correlate at the centennial level with the Greenland GISP2 ice core oxygen isotope ratios (Higginson *et al.*, 2003; Figure 4.32). Nitrogen isotopes are seen to respond to the glacial cycles over longer time periods because nitrogen is buffered by a multi-millennial year residence time in the oceans.

The close linkage between glacial–interglacial cycles (Dansgaard–Oeschger events) and Arabian Sea denitrification has implications for our understanding of global climate teleconnections. The δ^{15}N record demonstrates the high sensitivity of the Arabian Sea oxygen minimum zone to Northern Hemisphere climate change on short timescales. Altabet *et al.* (2002) demonstrated that Arabian Sea intermediate waters have been transformed, in a century or less, from an oxic state to a suboxic state that was more intense than that seen at present. Because denitrification is controlled by the biogenic productivity at shallow water depths, the δ^{15}N record implies large and very rapid changes in upwelling rates, which are largely controlled by summer monsoon winds.

Nitrogen isotopes were also employed by Tamburini *et al.* (2003) to examine changing monsoon-driven productivity in the southern and northern parts of the South China Sea using ODP Sites 1143 and 1144 respectively (Figure 3.7).

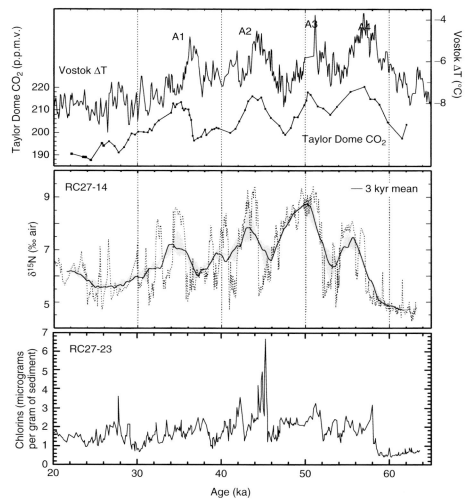

Figure 4.32 The 3 ky moving average of the $\delta^{15}N$ record offshore of Oman from Altabet *et al.* (2002) compared with Antarctic ice-core records of climate and atmospheric CO_2 (Indermühle *et al.*, 2000). The 61 ky envelope is shown shaded. Variations in Arabian Sea denitrification would alter the marine combined N inventory on timescales similar to its 3 ky residence time, in the absence of other effects. ΔT, the relative temperature change at Vostok derived from ice D/H ratios; A1–A4, Antarctic climate events. Reproduced with permission of Macmillan Magazines Ltd.

In this basin the links between monsoon strength and productivity are less well defined than in the Arabian Sea, but despite this a pattern was revealed showing linkage between Dansgaard–Oeschger events and productivity since 140 ka (Figure 4.33). It is nonetheless noteworthy that the patterns are not the same on the southern and northern margin of the basin, reflecting the asymmetry of

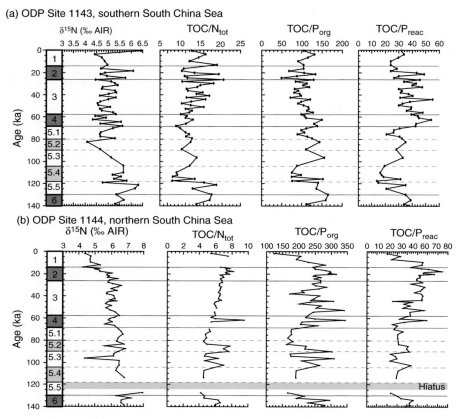

Figure 4.33 Nitrogen isotope record, TOC/N_{tot}, TOC/P_{org} and TOC/P_{reac} molar ratios from (a) ODP Site 1143 in the Dangerous Grounds and (b) ODP Site 1144 on the northern margin of the South China Sea (modified after Tamburini et al., 2003). Glacial stages are shown in the left-hand column with dark gray shading, interstadials in light gray and interglacials in white. Reprinted with permission of Elsevier B.V.

the upwelling patterns. At ODP Site 1143, in the southern South China Sea, $\delta^{15}N$ values correlate with calcite mass accumulation rates, with both being enhanced during interglacial periods. This is consistent with the fact that this drill site now lies close to the modern zone of summer monsoon upwelling offshore of southern Vietnam. In contrast, $\delta^{15}N$ values at ODP Site 1144 in the north are more constant, but then decrease to the present after the Last Glacial Maximum, ~ 20 ka, as the winter monsoon weakened.

4.9.2   Organic carbon and phosphorus

The ratio between total organic carbon (TOC) and total nitrogen (N_{tot}) values indicates that organic material at both ODP Sites 1143 and 1144 in the

South China Sea is composed of a mixture of degraded terrestrial and marine organic matter. Sediment recovered at ODP Site 1144 derived a significant amount of its nitrogen from inorganic sources. Nonetheless, there is no strong temporal change noted in TOC/N_{tot} values (Figure 4.33). High values of TOC/organic phosphorus (P_{org}) ratios are observed at ODP Site 1144, especially at around 100 ka and during the glacial Marine Isotopic Stages 4–2. These values can be used to infer the source of organic material because high values of TOC/P_{org} indicate derivation from terrestrial sources, while low values are associated with marine phytoplankton (Redfield *et al.*, 1963).

The high TOC/P_{org} values at ODP Site 1144 during Marine Isotopic Stages 4–2 (Figure 4.33) reflect the enhanced input of terrestrial material by eolian transport during strengthened winter monsoon intervals, and the preferential preservation of organic C over organic P, caused by low oxygen conditions in pore waters (Ingall and Jahnke, 1994). Tamburini *et al.* (2003) also noted that clay mineralogy on the northern margin of the sea indicated reduced chemical weathering during glacial periods, especially during Marine Isotopic Stages 4–2. Relatively high values of TOC/reactive phosphorus (P_{reac}) ratios parallel changes in TOC/P_{org} during Marine Isotopic Stages 2–4 at ODP Site 1144. This behavior may reflect either reduced input of phosphorus in detrital form to the basin during glacial period intervals, or regeneration and loss of phosphorus to the water column, for instance due to severe oxygen depletion. High P_{reac} and other detrital mass accumulation rates during glacial periods indicate an increased continental contribution to sedimentation, probably as a result of strong eolian sedimentation during an enhanced winter monsoon. These data were consistent with the suggestion of Wang *et al.* (1999a) that during glacial intervals of low sea-level and a weak summer monsoon, circulation was reduced in the South China Sea, leading in turn to increased oxygen deficiency in intermediate and bottom waters.

4.9.3   Opal and foraminifers

Paleoproductivity on the Oman margin of the Arabian Sea has been used to trace monsoon intensification in that region over long periods of geologic time (Anderson and Prell, 1993). Clemens *et al.* (1996) demonstrated orbital scale variability offshore of Oman through a high-resolution reconstruction of changing biogenic productivity at ODP Site 722. Figure 4.34 shows some of the proxies used to make the reconstruction, especially the proportion of the summer upwelling-related *G. bulloides*, together with opal contribution and mass accumulation rates that reflect the productivity of siliceous plankton in the near surface waters. Moreover, Clemens *et al.* (1996) extended the monsoon reconstruction far back into the geologic past to show short duration cyclicity extending not only through the Pleistocene, but also into the Pliocene.

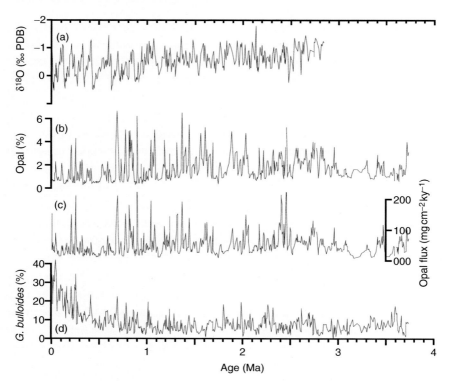

Figure 4.34 Proxy climate records from ODP Site 722 measured by Clemens *et al.*
(1996). (a) Global ice volume ($\delta^{18}O$, relative to the Pee Dee belemnite standard), larger
values indicate increased glaciation. (b, c and d) Biological productivity driven
by wind-induced upwelling during the summer monsoon (opal %, opal flux, and
G. bulloides %). Reproduced with permission of the American Association for the
Advancement of Science.

Because ice core records do not extend to the Pliocene, Clemens *et al.* (1996)
were forced to use the oxygen isotope record from the same drill core as a proxy
for ice volume. By doing this they were able to make a direct comparison
between monsoon paleoproductivity record and the intensity of Northern
Hemispheric Glaciation. Figure 4.34 shows some of the links between ice volume
and peaks and troughs in the biogenic production, highlighting the influence
that glaciation has had on monsoon strength and in turn on the oceanic biota.

Spectral analysis allows the relationships to be investigated in greater detail
in order to understand the overall controls to this system. While continental
aridity, tracked by eolian dust, was seen to be in phase with retreats and
advances in ice sheets, the same was not true for the opal record, which is not
in phase or even stationary relative to the precessional forcing function. Indeed,

Clemens *et al.* (1996) showed that the phase of monsoon-driven biogenic productivity measured by opal flux changed from $-225°$ to $-140°$ over the course of the record since 3.5 Ma. It appeared that the rate of change was linked to rates of change in the intensity of Northern Hemispheric Glaciation. Changes at 2.6 Ma correlate with the start of major ice sheet development in the northern hemisphere. At around 1.2 Ma the ice sheets started to evolve on a 100 ky, eccentricity-driven timescale and at 0.6 Ma this control became even stronger (Wang and Abelmann, 2002). As these changes occurred the intensity of the monsoon shifted to be more in phase with the timing of ice minimum. Spectral analysis suggests that before ~1.7 Ma opal productivity was the most robust summer monsoon proxy rather than the eolian dust record.

4.10   The Early Holocene monsoon

4.10.1   *Climate modeling*

Although representing only the most recent time period the superior age control offered by the geological record spanning the transition from the Last Glacial Maximum (~20 ka) to the present allows the short-term evolution of the Asian monsoon to be accurately reconstructed and the controls on its intensity isolated. Moreover, because the climatic and geologic conditions during the Younger Dryas, Bølling-Allerød and Early Holocene are similar to the modern period, it is possible to make relatively realistic climate models for this time period in order to constrain controls. Bush (2004) used numerical modeling methods to understand the changes in summer monsoon precipitation between the Last Glacial Maximum, the Early–Mid Holocene and the present day. In this study a modification of the General Circulation Model (GCM) was used that gave a spatial resolution of 3.75° in longitude and 2.25° in latitude. Temperature, wind and precipitation predictions were made for the different stages of deglaciation based on what is known about changing atmospheric CO_2 concentrations, the exposure of continental shelves due to lower sea levels (Fairbanks, 1989) and the aerial extent of ice sheets (Peltier, 1994). These latter two factors are important because they affect the albedo of the continents and thus the temperature differences between land and ocean that are the root cause of the Asian monsoon.

The Bush (2004) GCM model showed that the Asian monsoon since 20 ka has been controlled by orbitally modulated insolation, by the changing temperature in the Western Pacific Warm Pool and by exposure of the Sunda Shelf. The model predicts that the western Pacific region would have been much drier during the Last Glacial Maximum as a result of the exposure of the Sunda Shelf to subaerial conditions. The Mid Holocene model predicts lower rainfall in Central Asia, but heavier precipitation along the Himalayan front, owing

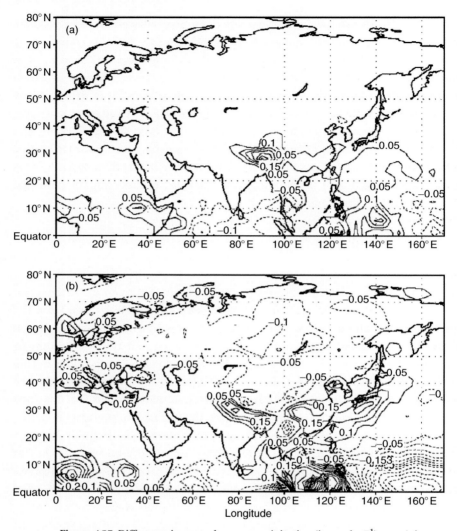

Figure 4.35 Difference in annual mean precipitation (in cm day^{-1}) over Asia calculated by modeling methods by Bush (2004) during (a) the Mid Holocene and (b) at the LGM. The contour interval in both panels is 0.05 cm day^{-1}. Images reproduced with permission of Taylor & Francis Journals.

to changes in the direction of monsoon winds. Stronger summer monsoon winds enhanced evaporation over the Arabian Sea bringing that moisture to the Himalayas (Figure 4.35). Some of this extra precipitation accumulated as snow, with the eastern Himalayas predicted to have been an area for accumulation during the Last Glacial Maximum. In the Mid Holocene simulation, annual mean surface temperatures across northern Asia are projected to have been warmer than present and to have exhibited a stronger seasonal cycle, consistent with a stronger obliquity-driven forcing. In contrast, temperatures were cooler across South Asia, owing to a warmer Western Pacific Warm Pool. Central Asia is modeled to have been slightly drier, although models predicting climate based only on atmospheric changes caused by insolation indicate that this drying is driven by solar processes and is not linked to sea-surface temperatures. The dry conditions in Central Asia are linked strongly to the extent of the Fennoscandian ice sheet.

Changes in monsoon intensity since the Last Glacial Maximum are known to have effects that extend beyond the Asian mainland, as a result of teleconnections with the El Niño Southern Oscillation (ENSO). The nature of monsoon–ENSO interactions has recently been revealed by coral records from Indonesia, which were used to assess the changing sea-surface temperatures in this region since the Last Glacial Maximum (Abram *et al.*, 2007). They were also used to trace evolution of the Indian Ocean Dipole, an oceanographic system that involves a reversal of the sea-surface temperature gradient and winds across the equatorial Indian Ocean on an eight to nine year timescale (Saji *et al.*, 1999). As well as affecting the Indian monsoon, the Indian Ocean Dipole causes periodic drought conditions in Indonesia. This study showed that Indian Ocean Dipole events during the Mid Holocene (~6.5 ka), when speleothem records of the summer monsoon indicate a peak in strength, were characterized by a longer duration of strong surface ocean cooling, together with droughts that peaked later than those expected by ENSO forcing alone.

4.10.2   Peat bog records

Modeling studies such as Bush (2004) are important in allowing the sensitivity of the monsoon strength to different forcing controls to be assessed. Nonetheless, unless they are grounded in observations their use is necessarily limited. The cave and ice records discussed above provide important constraints on the Holocene monsoon climate around Asia. In addition, recent work on deposits from peat bogs allows the temporal variation in monsoon strength to be reconstructed during the Early Holocene. Carbon isotopes can be used to assess changing flora and the intensity of the summer monsoon using the peat. Hong *et al.* (2001, 2003) have demonstrated that the stable carbon isotope composition of plant cellulose extracted for peat beds can be exploited as a proxy for the Asian summer monsoon activity. Low $\delta^{13}C$ values are interpreted to indicate strong activity of the summer monsoon (i.e., a wetter, warmer climate), while higher $\delta^{13}C$ values indicate a weaker summer monsoon, with a dry cold climate.

Hong *et al.* (2005) presented a carbon isotope record spanning the last 12 ky (i.e., since the Younger Dryas) for a bog peat from NE China, complementing that published by Hong *et al.* (2001, 2003) from SE Tibet. These two locations are

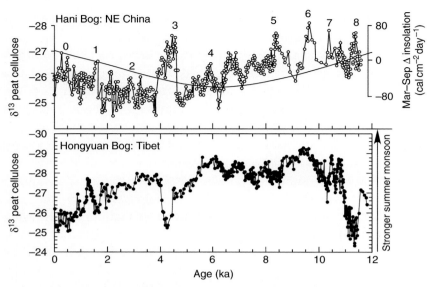

Figure 4.36 δ^{13}C values measured at Hani Bog in NE China (Hong et al., 2005), and at Hongyuan Bog in Tibet (Hong et al., 2003). Numbers indicate Bond ice rafting events, with 0 being the Little Ice Age. The curved line shows the insolation difference between March and September at the equator.

important because each lies within the sphere of influence of the South and East Asian monsoon systems, and allows their development at these short timescales to be compared. Figure 4.36 shows the rather surprising result that the δ^{13}C values for cellulose taken from the peat vary in an inverse way between the two regions. Over the total time of the climate reconstruction δ^{13}C values at each area vary in the opposite directions, e.g., since 4.0 ka δ^{13}C values have risen at Hani Bog in NE China, while falling at Hongyuan Bog in Tibet. This long-term variability may be tied to the variations in insolation over that interval. The South Asian summer monsoon is weak during the Younger Dryas around 11.5 ka, but then strengthens to be strong from 10.0 to 5.5 ka. In contrast, these bog records indicate a strong East Asian summer monsoon during the Younger Dryas, which then weakens until ~4.0 ka. This result may be suspect because the speleothem records from Hulu Cave appear to show a weak summer monsoon during the Younger Dryas (Zhao et al., 2003). It is possible that the Hani Bog record is affected by other processes, as well as the Asian monsoon during this period. The major isotope event at ~4.2 ka, which corresponds in time to Bond ice-rafting event 3 from the North Atlantic (Bond et al., 2001), shows very clearly the opposing polarity of the two monsoon systems, which respond in opposite fashion to the same forcing control.

4.10.3   El Niño Southern Oscillation effects

Hong *et al.* (2005) suggested that El Niño-like conditions are coincident with stronger East Asian summer monsoons, weaker Indian summer monsoons, and ice drift events in the North Atlantic at both orbital and millennial timescales. The orbital-scale inverse relationship between the two monsoons indicates the occurrence of a long-term El Niño Southern Oscillation (ENSO)-like pattern at least since 12 ka. This implies that the monsoon systems are sensitive to insolation forcing through ENSO variations, as also demonstrated by Abram *et al.* (2007) in the Indian Ocean. Hong *et al.* (2005) identified nine inverse phase variations since 12 ka, and suggested that the nine El Niño-like periods were closely correlated to the Bond ice drifting events at high northern latitudes in the Atlantic Ocean. In every case when the abrupt ice-rafted debris events occurred in the North Atlantic, the inverse phase relationship was established, and an El Niño-like pattern occurred in the tropical Pacific correspondingly. The Bond event at 4.2 ka is especially noteworthy in its intensity and opposing effects in South and East Asia. It is possible that large-scale addition of melt water pulses into the Atlantic during Bond events disrupted the normal global thermohaline circulation. If this became less efficient then the temperature contrast between the Western Pacific Warm Pool and the rest of the Pacific would be enhanced and favor the onset of El Niño conditions.

4.10.4   Links to ice volume and CH$_4$ levels

The peat reconstructions can be cross checked with other high resolution data for the Early Holocene, one of the best being the ice core records from the Guliya Ice Cap of western Tibet discussed above (Thompson *et al.*, 1997). Figure 4.37 shows the changing strength of the summer monsoon rains at Guliya, which increased significantly during the Bølling-Allerød, fell again during the Younger Dryas, and then rose again in the Preboreal part of the Early Holocene. What is interesting to note is that by the middle Holocene $\delta^{18}O$ values for Guliya ice had fallen again, while they remain steady and high in Greenland (GISP2 core, Alley *et al.*, 1993). In this respect the Guliya ice shows closer correspondence with the East rather than the South Asian monsoon. This diagram also shows a weak coupling between the Greenland and monsoon systems, but a closer correlation between monsoon strength and atmospheric CH$_4$ levels (Thompson *et al.*, 1997). Intensification of the monsoon over Asia during the Early Holocene is inferred primarily from lake levels and lake sediments, which indicate that the climate of central China was warmer and wetter in the Early–Mid Holocene (Fang, 1991; Gasse *et al.*, 1991; Herzschuh, 2006). This contrasts with the Hani peat record and suggests that NE China may have experienced a contrasting Holocene monsoon history to the rest of Asia. The Guliya $\delta^{18}O$ record

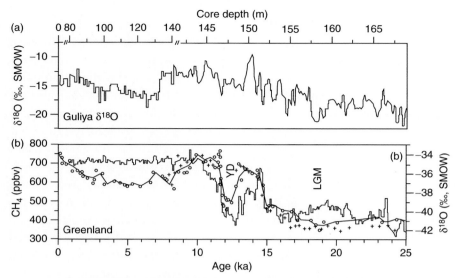

Figure 4.37 The Guliya $\delta^{18}O$ record covering the last deglaciation is compared with the (b) Greenland CH4 (GRIP, 1; GISP2, E) and $\delta^{18}O$ profiles (GISP2, thin line). 500-year averages of GISP2 CH4 are also shown (thick line). Modified from Thompson *et al.* (1997). Reproduced with permission of American Association for the Advancement of Science.

also suggests that higher levels of atmospheric CH_4 in the Early Holocene correlate with warmer, moister conditions in the subtropics. This in turn supports the idea of a low-latitude methane source being an important control on global climate at that time.

4.10.5   Solar forcing

The centennial-scale development of the Early Holocene monsoon has been reconstructed in detail in the Arabian Sea area using a combination of speleothem records from northern Oman (Hoti Cave; Neff *et al.*, 2001) and a deep-sea core from offshore Somalia (Jung *et al.*, 2004). Both these records show evidence for dramatic short-term change in the monsoonal climate, as manifest from the changing $\delta^{18}O$ values of the speleothems and the abundance of the upwelling-loving *G. bulloides* foraminifers respectively (Figure 4.38). Naqvi and Fairbanks (1996) used another upwelling record from the Red Sea to suggest a lag between monsoon intensification between the Last Glacial Maximum and 15 ka, but no lag after that time. The fact that high-magnitude changes affected both the cave precipitation records and the marine temperature record offshore of Somalia indicates that insolation was controlling both aspects of the monsoon system in parallel, despite the large distance between the two areas. Neff *et al.* (2001)

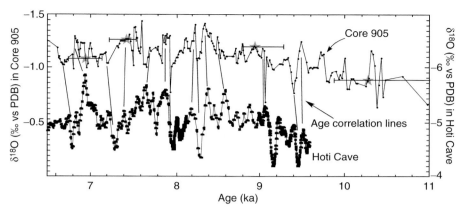

Figure 4.38 Stable O-isotope records of *G. bulloides* from Core 905 off Somalia (Jung *et al.*, 2002) and from a stalagmite from Hoti Cave (Neff *et al.*, 2001) based on their individual age models. Correlation lines show the main tie points used to adjust the age model of Core 905 to that of Hoti Cave. A 40-year moving averaging filter was applied to both records. Asterisks mark AMS[14]C-dates including two sigma uncertainty ranges. Reprinted with permission of the American Geophysical Union.

inferred that these two aspects could be reconciled if insolation maxima resulted in the Intertropical Convergence Zone being displaced northwards, driving up precipitation in Oman.

At the same time, sea water offshore of Somalia was calculated to have been 2–3 °C warmer, which Jung *et al.* (2002) noted was a much larger change that might have been expected for what were modest changes in the insolation, not least because of the large heat capacity of the oceans. These workers suggested that if the Intertropical Convergence Zone moves north then this would have reduced the gradient of atmospheric circulation cells, weakening the SW monsoon and thus reducing the intensity of upwelling of cooler, deep-sourced waters. However, on longer timescales insolation maxima are associated with strong summer monsoons and more upwelling. It is possible that this is occurring offshore of Somalia too, with higher sea-surface temperatures caused by more intense mixing of cold upwelling waters with warm surface waters.

Neff *et al.* (2001) used a spectral analysis method to show that the periodicities in the speleothem record matched those from tree-ring records, which are known to be linked to insolation. Variations in summer monsoon precipitation were noted over time periods of 779, 205, 134 and 87 years. The periodicities observed in the high-resolution speleothem record over the time interval 7.9–8.3 ka are similar to the cycles of 24, 7.5 and 3.2 years that had already been noted for the North Atlantic Oscillation (e.g., Cook *et al.*, 1998), suggesting that

at these shorter timescales the North Atlantic climate system or its controls
might also be a cause of changes in the amount of monsoon rainfall in the Hoti
cave area of eastern Arabia.

4.11   Mid–Late Holocene monsoon

The spatial and temporal evolution of the monsoon during the Mid to
Late Holocene will be discussed further in Chapter 6, together with its implications for mankind. Here, however, we highlight some studies that relate to the
processes controlling millennial-scale variability. The Holocene monsoon has
been investigated by a study of a series of cores taken across the northern
Arabian Sea (Dahl and Oppo, 2006). The $\delta^{18}O$ and Mg/Ca analyses were made
from planktonic foraminifers (*Globigerinoides ruber*) picked from the cores in
order to constrain sea-surface temperature and salinity for four time intervals
since the Last Glacial Maximum (0 ka, 8 ka, 15 ka and 20 ka). This is a powerful
analytical combination because foraminiferal Mg/Ca is controlled by sea-surface
temperatures (Anand *et al.*, 2003), while $\delta^{18}O$ values reflect both salinity and
temperature. Salinity can thus be calculated from $\delta^{18}O$ once temperature has
been determined.

Dahl and Oppo (2006) showed that sea-surface temperatures were colder at
8, 15 and 20 ka than at present for the majority of sites (Figure 4.39). This is
important because although monsoon-driven upwelling brings cold, nutrient-
rich waters to the surface offshore of Oman, this process should most strongly
affect foraminifers living close to the coast and not further offshore. A more
regional cooling suggests a regional or global climatic process rather than the
monsoon. The 20 ka and 15 ka time slices display average negative temperature
anomalies of 2.5–3.5 °C attributable, in part, to the influences of the glacial
climate at this time. Crucially, the lack of a gradient in sea-surface temperatures
from east to west across the Arabian Sea at 15 and 20 ka suggests a weak summer
monsoon and a stronger winter monsoon (Dahl and Oppo, 2006). Changes in
$\delta^{18}O$ values in the foraminifers that are smaller than the $\delta^{18}O$ signal due to
global ice volume reflect decreased evaporation and increased winter monsoon
mixing during the Early and Mid Holocene. Sea-surface temperatures throughout the Arabian Sea were still cooler than present by an average of 1.4 °C at 8 ka.
Cool temperatures and lower $\delta^{18}O$ values at 8 ka can be interpreted to reflect a
stronger monsoon, driving increased continental run-off and precipitation, than
either the modern summer or winter monsoon.

Further detail on the paleoceanography of the Arabian Sea during the
Holocene was provided by analysis of rapidly accumulated, laminated sediment
from ODP Site 723 and Core RC2730 offshore of Oman (Gupta *et al.*, 2003)

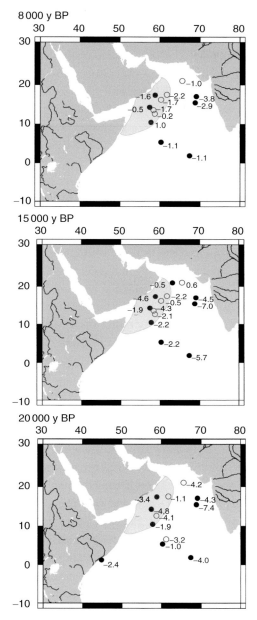

Figure 4.39 Maps showing temperature anomalies relative to the present day for surface waters in the northern Arabian Sea calculated by $\delta^{18}O$ and Mg/Ca analyses from planktonic foraminifers (Dahl and Oppo, 2006). Shading indicates the region currently affected by summer monsoon upwelling. Open circles represent cores for which anomalies were calculated using the time slice sample and the nearest modern core top sample. Reprinted with permission of the American Geophysical Union.

(Figure 4.3). This study focussed on changes in monsoon intensity since the start of the Holocene, after the major climatic fluctuations related to the end of the Younger Dryas had subsided. Gupta *et al.* (2003) pointed out that despite the relatively moderate changes in ice volume since that time monsoonal strength, traced by abundance of *G. bulloides* and $\delta^{18}O$ values of these foraminifers (both proxies for monsoon-driven upwelling strength) has changed greatly on centennial timescales. In detail those periods when the monsoon winds and coastal upwelling were weak correlated with times of greater cold in the North Atlantic, as recorded in the GISP2 ice core (Figure 4.40).

The strength of summer monsoon-induced upwelling has decreased since 12 ka, in parallel with decreasing insolation. However, when this long-term trend is removed, the record of upwelling still shows centennial-scale variations that correlate with ice-rafting Bond events in the North Atlantic. This correlation between North Atlantic temperatures and the monsoon intensity extended even to the relatively modest climatic fluctuations of the Medieval Warm Period and the Little Ice Age. The driver of these fluctuations is likely short-term variations in insolation, although Gupta *et al.* (2003) did raise the possibility that variability internal to the North Atlantic thermohaline circulation system could have driven the ice rafting events and consequently also the strength of the SW Asian monsoon.

It has been suggested that the temperature of the North Atlantic affects the monsoon through control of the snow cover in Asia and especially over the Tibetan Plateau. More snow on the plateau in spring or early summer uses up all the sun's heating because it has to be melted and evaporated before the land can warm. As a result winters with heavy snow cover cause weaker monsoons in the following summer. The North Atlantic is able to influence the monsoon system because Tibet is downwind of this water mass and is cooled when this cools.

Centennial and decadal scale variations in monsoon strength during the Holocene have been reconstructed offshore of Pakistan using laminated sediments deposited in anoxic conditions within the oxygen minimum zone (Doose-Rolinski *et al.*, 2001). In these regions, a lack of bioturbation allows very thin beds to accumulate and be preserved, resulting in very detailed climate reconstructions. Oxygen isotopes were measured from planktonic foraminifers extracted from the sediments, together with alkenone data from the organic material that can be independently used to constrain the sea-surface temperatures.

Centennial and decadal scale changes in the sea-surface temperatures offshore of Pakistan were noted to range up to 3 °C of magnitude, which were interpreted as reflecting the changing strength of the monsoon, despite the relative constancy in the intensity of Northern Hemispheric Glaciation since 5 ka. Figure 4.41 shows that these variations were paralleled by changes

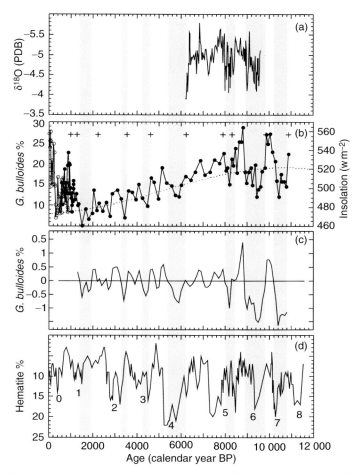

Figure 4.40 South-west monsoon proxy record from the Arabian Sea ODP Site 723A and box core RC2730 from Gupta *et al.* (2003) combined with Oman cave stalagmite $\delta^{18}O$ and North Atlantic haematite percentage. Time series of (a) cave stalagmite $\delta^{18}O$ from Neff *et al.* (2001), (b) *G. bulloides* percentage in ODP Hole 723A (filled circles) and box core RC2730 (open circles) from the Arabian Sea, and July insolation at 65° N (shown by dotted line; radiocarbon-dated intervals shown by crosses), (c) change in *G. bulloides* percentage (normalized by removing the trend related to insolation) and (d) haematite percentage in core MC52-VM29-191 from the North Atlantic that can be used as a tracer of ice-rafting events 0–8 from Bond *et al.* (2001). The vertical gray bars indicate intervals of weak Asian SW monsoon. Reproduced with permission of Macmillan Magazines Ltd.

in the thickness of the annual varves in these laminated sediments. Because varve thickness represents the amount of sediment deposited in each year it follows that these should record the amount of clay carried by run-off from the continents. As might be expected, this study showed that periods of thick varve

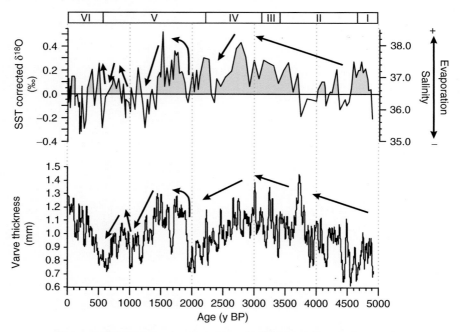

Figure 4.41 Reconstruction from Doose-Rolinski *et al.* (2001) of sea-surface temperature (SST) corrected δ¹⁸O records from core 39KG-56KA with varve thickness records from the Pakistan margin (von Rad *et al.*, 1999). Reprinted with permission of the American Geophysical Union.

deposition coincided with times of lower salinity (as calculated from the δ¹⁸O values and the alkenone data). Together these indicate more run-off of fresh, sediment-bearing fluvial waters into the Arabian Sea at times when the global climate was warmer. When compared with the present salinity of the Arabian Sea the Pakistan margin record since 5 ka shows generally lower salinities and thus stronger summer monsoons than has been typical since 1500 y BP. The Late Holocene monsoon is discussed in greater detail in Chapter 6, where its direct influence on human development is investigated.

4.12   Summary

A host of different proxies from both terrestrial and marine settings in both East and South Asia show strong evidence for millennial, centennial and shorter scale variations in monsoon strength during the Neogene and especially since the onset of Northern Hemispheric Glaciation ~2.7 Ma. Most of the evidence comes from high-resolution records dating back ~100–200 ky that show a close correspondence between stronger summer monsoons and warmer periods during the glacial–interglacial Dansgaard–Oeschger cycles. Because the monsoon strength varies on the same timescales as the Northern Hemispheric Glaciations it is inferred that the two systems are both being ultimately controlled by the changing intensity of insolation, which controls seasonality, as well as the total amount of solar energy available to warm the planet. In recent geologic times the Asian monsoon is mostly linked to the precessional cycle of 21 ky duration.

Records from the Loess Plateau indicate that prior to 1.6 Ma the Asian winter monsoon was variable, but not in phase with the dominant 41 ky cycle of the Northern Hemispheric Glaciation. The 100 ky eccentricity cycle that is the primary pattern of Northern Hemispheric Glaciation is more important in the monsoon variations after 1.0 Ma, especially in controlling the East Asian Winter monsoon intensity recorded in the Loess Plateau and North Pacific Ocean. The summer monsoon, which governs the continental aridity, is more closely tied to the 21 ky precessional cycle.

While there does appear to be a close correlation between the monsoon and Northern Hemispheric Glaciation it is noteworthy that the sharp warming phases at the start of the Bølling-Allerød and at the end of the Younger Dryas in the North Atlantic are not mirrored perfectly in the monsoon system, where the climatic transition to a wetter summer monsoon was much more gradual. However, climatic shifts to colder global conditions and stronger winter monsoons seemed to be rapid and synchronous.

Temporal changes in the strength of the Asian monsoon represent a dramatic example of the widely occurring and apparent synchronous climatic events seen in different oceans, and which coincide with the Dansgaard–Oeschger climatic variations seen in the North Atlantic. The cycles in monsoon strength on these orbital timescales may be caused by changes in atmospheric circulation and ocean ice cover (Mayewski *et al.*, 1994), atmospheric teleconnections between the North Atlantic and Asia (Hostetler *et al.*, 1999; Porter and An, 1995), the changing abundance of dust aerosols in the atmosphere over Asia (Overpeck *et al.*, 1996b), tropical heat budgets (McIntyre and Molfino, 1996; Cane, 1998) or equatorial wind stress (Klinck and Smith, 1993). Links have also been made between Antarctic climate and monsoon intensity. Several processes have the potential to link the climates in Asia and the North Atlantic and drive changes in monsoon intensity. Precisely how these teleconnections work is not yet clear.

The CH_4 records from Antarctic ice cores, compared with Tibetan ice cores, indicate that global CH_4 levels and the monsoon climatic cycle are linked (Thompson *et al.*, 1997), although it is possible that they are both forced by some other system, likely insolation. The linkage between monsoon intensity and CH_4 levels in the Holocene when global ice volumes were relatively stable, shows that the monsoon is not mere controlled by a direct and simple coupling to the intensity of glaciation.

The current interpretation of the links between Northern Hemispheric glaciation and monsoon strength is that changes in the climate in the North Atlantic drive changes in the global ocean or atmospheric circulation, and this in turn drives changes in the Asian monsoon. Heinrich iceberg calving events preceded the end of glacial periods and resulted in major re-organization in Atlantic circulation patterns that may have influenced global ocean circulation and the land–sea temperature differences. Alternatively, colder air over Greenland may have driven a hemisphere-wide atmospheric cooling, intensifying the Siberian High and thus strengthening the winter monsoon. It is however worth considering whether the reverse might be the case. Wang *et al.* (2003c) have argued that monsoon-driven processes in the low latitudes of Asia and the western Pacific control the storage or release of organic carbon from marine sedimentary reservoirs. Periods of high marine $\delta^{13}C$ seem to precede major falls in global temperatures and the expansion of ice sheets. Thus the Asian monsoon, modulated by insolation, might in part be controlling the strength of Northern Hemispheric Glaciation.

The monsoon can have an effect on global climate as a result of the large variations in the amount of water vapor introduced into the atmosphere over large parts of South and East Asia and which must influence the conditions that produce the greenhouse effect (Raval and Ramanathan, 1989). This effect mostly seems to work as a negative feedback. During warm, wet interglacial periods, high rates of transport of water vapor from ocean to continental areas increased continental snow and ice accumulation (Kapsner *et al.*, 1995) and caused falling sea level and a weaker summer monsoon. In contrast, during cold, dry and dusty glacial times a more balanced ocean–continent vapor budget and a stable sea level, resulting in an enhanced monsoon, is predicted. The links between the Asian monsoon and other parts of the global climate system are still being explored, yet it is clear that, because of its shear size and intensity, the monsoon is a crucial player in the network of climatic systems that are interlinked through teleconnections, operating through both oceanic and atmospheric processes.

5

Erosional impact of the Asian monsoon

The changing strength of the Asian monsoon described in preceding chapters has had a major effect on the continental climate of Asia, and the oceanography of the surrounding seas. The monsoon is partially controlled over time spans >1 My by the tectonic evolution of the solid Earth. However, in turn, the monsoon has had an influence on the geology and tectonics across South and East Asia. This is for the simple reason that climate, and especially rainfall, is a major factor controlling the erosion of the continental crust, and this in turn affects how mountain belts are exhumed and uplifted. Because of the huge area of the monsoon-affected regions and the dominance of the Himalayas and Tibet in the global topography, continental erosion driven by the monsoon can have far-reaching consequences, not only for the region, but also for the global ocean through run-off via some of the world's largest rivers. In this chapter we examine the evolving erosion of Asia in the context of the monsoon and assess how the monsoon may have controlled the tectonics of the Asian mountain ranges.

5.1   Monsoon and oceanic strontium

The potentially global importance of monsoon-driven erosion has been recognized for some time now via the evolving Sr isotope character of the oceans (Figure 5.1; Rea 1992). The isotope character of sea water shows a steady, coherent evolution to more radiogenic values since the Eocene ($\sim$40 Ma). Because the curve is necessarily a global integration it reflects input from three main sources, from oceanic hydrothermal sources, from dissolution of marine carbonates and from continental run-off. The last of these has the correct end-member composition to shift the global sea water balance in the way observed since 40 Ma. Palmer and Edmond (1989) recognized that it was weathering of metamorphic

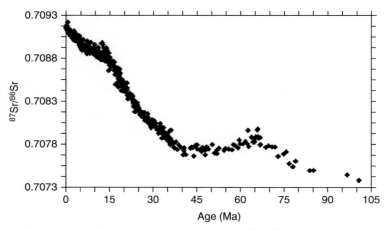

Figure 5.1 Late Cretaceous and Cenozoic seawater $^{87}Sr/^{86}Sr$ curve compiled from multiple sources from the study of Rea (1992), showing the consistent shift in values to higher ratios since 40 Ma, mostly related to weathering of the Himalayas and Tibet via run-off in the Brahmaputra River. Reprinted with permission of the American Geophysical Union.

carbonate rocks in the Himalayas, and especially in the Ganges-Brahmaputra drainage, that was the largest source of radiogenic Sr to the ocean. Because this region is also the location for maximum summer monsoon rains it follows that monsoon strength must play a part in controlling the flux and thus maintaining the $^{87}Sr/^{86}Sr$ value of oceans.

Figure 5.1 shows that since 40 Ma the gradient of the curves steepened slightly around 20–25 Ma and again in the last 3 Ma. The $^{87}Sr/^{86}Sr$ values continued to climb, but flattened significantly around 16 Ma, suggesting a reduced flux from the Ganges-Brahmaputra drainage after that time. It is hard to correlate these isotopic events with tectonism in the Himalayas, although 20–25 Ma does appear to be a time of major exhumation in the Greater Himalayas (e.g., Hodges and Silverberg, 1988; Treloar *et al.*, 1989; Guillot *et al.*, 1994; Hodges *et al.*, 1996; Walker *et al.*, 2001; Godin *et al.*, 2006). Changes in the $^{87}Sr/^{86}Sr$ slope at 16 and 3 Ma do not correlate with major known tectonic events and may reflect climatic episodes. Rea (1992) compared the $^{87}Sr/^{86}Sr$ history with what was then known of the evolving erosion of Asia and concluded that the two must be significantly decoupled, despite the chemical arguments linking the global $^{87}Sr/^{86}Sr$ value to flux from the Bengal delta.

5.2   Reconstructing erosion records

Whether the global oceanic $^{87}Sr/^{86}Sr$ is truly detached from erosion in the monsoon catchment can now be better assessed using improved erosion

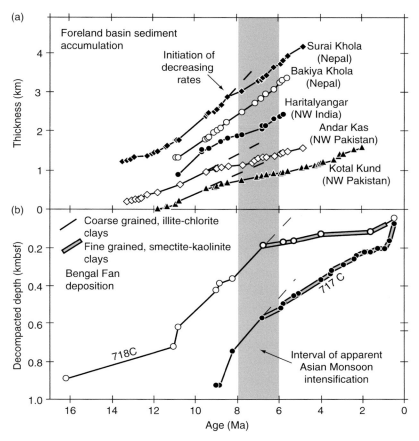

Figure 5.2 (a) Evolving rates of sediment accumulation in the Himalayan foreland basin, modified after Burbank *et al.* (1993), showing the slowing of rates after ∼9 Ma. (b) The depth versus time in two ODP Sites 718 and 717 on the distal fan, after correction for porosity reduction with burial. These show slowing rates after 8 Ma, consistent with a reduced sediment flux to the deep sea fan, correlated by Burbank *et al.* (1993) with monsoon intensification. Reproduced with permission of Macmillan Magazines Ltd.

records from across Asia. Figure 5.2 shows Burbank *et al.* (1993)'s reconstruction of sediment accumulation in the central and eastern Himalayas, showing a sharp drop in the rate of clastic accumulation on the Bengal Fan and in some parts of the Himalayan foreland basin after 8 Ma. Assuming 8 Ma to be the time of initial monsoon strengthening and thus heavier summer rains Burbank *et al.* (1993) concluded that a stronger monsoon resulted in less erosion in the Ganges-Brahmaputra catchment. Although this seemed counter-intuitive, these authors reasoned that the warming and moistening of the air over South Asia resulted in reduction of erosive glaciers and increased vegetation that reduced run-off and soil erosion from mountain slopes. While these arguments are not illogical it is nonetheless noteworthy that those regions of the Himalayas where annual precipitation is now heaviest (i.e., in the eastern Himalaya) are also now the areas where modern erosion rates are at a maximum (Galy and France-Lanord, 2001) and the river clastic sediment loads are among the highest in the world (Milliman and Syvitski, 1992). Recent river and thermochronology work from Taiwan and the Cascades (e.g., Dadson *et al.*, 2003; Reiners *et al.*, 2003) also shows a first-order correlation between erosion rates and precipitation that now make the proposed inverse link between erosion and monsoon strength seem suspicious. Erosion was also seen to increase in those parts of the Himalayas that are normally arid when stronger than normal monsoons penetrate beyond the topographic rainfall barrier (Bookhagen *et al.*, 2005b).

What is important to remember when considering the accumulation rates reconstructed by Burbank *et al.* (1993) is that these charts represent preservation at a single point, either at an Ocean Drilling Program drill site on the distal Bengal Fan, or along a section within the foreland. It is not clear whether these points are representative of the total mass flux in the river system at that time. This is especially true in the Himalayan foreland where accommodation space is necessarily limited compared with sediment flux, so that preserved sediment thicknesses may equate more with the degree of flexurally driven subsidence than with sediment supply. In practice the Himalayan foreland basin is always full so the preserved thickness in any time interval cannot reflect the rate of erosion in the sources. The marine record may be a more reliable proxy for continental erosion, especially in the Bengal Fan where the sediment supply is very high and most of the accommodation space lies offshore. Nonetheless, even in this setting the deposition at any one point may more reflect the lateral migration of channel–levee complexes across the fan (Schwenk *et al.*, 2003). In addition, sea-level variations can result in large changes in sediment depocenter, with more sediment being preserved on the delta and shelf during sea-level high stands, but being preferentially redeposited into deeper water as the sea level falls and reaches low-stands (e.g., Vail *et al.*, 1977; Haq *et al.*, 1987). As such it is only by summing all, or at least most of the offshore sediment record that a reliable erosional proxy can be reconstructed.

To derive a more representative sediment flux reconstruction the masses of eroded rock in each major depocenter must be estimated. In the case of the Ganges-Brahmaputra drainage this lies offshore in the Bengal Fan, where sediment thicknesses reach 22 km (Figure 2.3; Curray and Moore, 1971; Curray, 1994; Curray *et al.*, 2003). Although several other workers have attempted to quantify the flow of eroded rock from Asia into the oceans during the Cenozoic (e.g., Davies *et al.*, 1995; Métivier *et al.*, 1999) these efforts have often been hampered by the use of one-dimensional borehole data, often in distal areas of the sediment mass, or by industrial boreholes, located on continental shelves where accommodation space limits the ability of the basin to record the mass flux accurately. Nonetheless, accounting for the marine sediment bodies is crucial because this is where the bulk of the sediment resides.

5.3   Reconstructing exhumation

Although the volumes of sediment in the Himalayan foreland basin cannot tell us much about erosion rates caused by either the monsoon or tectonic activity in the sources, the mineral grains in the eroded sediments can be employed to reconstruct source exhumation rates. An example of this approach is shown in Figure 5.3, which plots the cooling of age of detrital mica grains in sandstones from the Himalayan foreland basin against their depositional ages in order to quantify exhumation rates (White et al., 2002). The Ar-Ar ages in micas are set when a rock cools through ~350 °C (Purdy and Jäger, 1976; Mattinson, 1978), equivalent to ~11 km depth in most continental environments. Provided the sediment is not subsequently deeply buried and the mica ages re-set, it is possible to assess exhumation rates through comparison of depositional ages and cooling ages. When these ages are close to one another then exhumation must have been rapid.

White et al. (2002) showed that in sediments deposited between 21 and 17 Ma (Early Miocene) the mica cooling ages in the western Himalayan foreland basin spanned a wide range of ages, though dominated by erosion of sources that were metamorphosed after the India–Asia collision (<50 Ma). Critically, some grain cooling ages approached the depositional age indicating that some sources were exhuming very rapidly at that time. However, after 17 Ma this pattern changed, with the youngest grains being at least 10 My older than the depositional age and with a number of older ages being recognized too. White et al. (2002) interpreted this shift to reflect a slowing of exhumation in the Greater Himalayas and increased erosion in the Lesser Himalayas, as the active thrust front propagated south. It is however worth considering whether the slowing of exhumation at 17 Ma might be linked to a change in monsoon climate, with slower exhumation being driven by less precipitation in the Greater Himalayas. The climate records from the Arabian and South China Seas indicate a short dry period at 16–17 (Figures 3.14 and 3.34), but then renewed wet conditions in the Middle Miocene (15–10 Ma). As discussed below, exhumation continued at a rapid pace during this time in the central and eastern Himalayas. Nonetheless, a change in monsoon strength and location appears to have triggered an erosional response that is preserved in the Himalayan foreland basin.

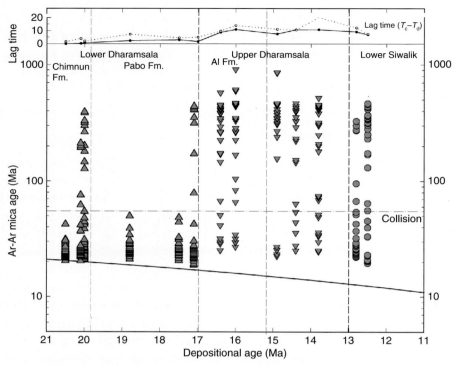

Figure 5.3 Measurement of changing source exhumation rates by White *et al.* (2002) from the Himalayan foreland basin. The upper panel shows the lag time, which is the difference between the youngest mica cooling age (T_c) and its depositional age (T_d), plotted against depositional age (solid line). Also shown is the modal mica lag time (dashed line). Note that the lag time is within 1–2 My for the Lower Dharamsala Formation and increases to 7–8 My for the Upper Dharamsala and Lower Siwalik, i.e., after 17 Ma. The lower panel shows single-grain mica ages plotted against time. The dash-dot line represents the time of India–Asia collision at ~55 Ma and marks the division between 'Himalayan' (<55 Ma) and 'pre-Himalayan' (>55 Ma) mica ages. The 1:1 line (solid) where mineral cooling age = depositional age is shown. Note the sharp decrease in Himalayan ages with an increase in pre-Himalayan ages after 17 Ma. Reprinted with permission of Elsevier B.V.

5.4   Estimating marine sediment budgets

Reliable estimates of erosion rates can be determined through geophysical seismic reflection surveying of the thickest sediment bodies, but only when these surveys can be tied to drill sites where microfossil evidence can be used to impose an age model on the deposits. When the density of the seismic data allows, it is possible to define volumes of eroded sediments that may be mass balanced with estimates of depth and area of erosion from onshore. One of the best surveyed regions in the Asian marginal seas is the Indus submarine fan in the Arabian Sea (Figure 2.3). Although some of the sedimentary flux from the Indus River may have flowed west into the Gulf of Oman during the Paleogene the vast bulk of the clastic flux from the Indus River has ponded east of the Murray Ridge since 20 Ma, when the Owen and Murray Ridges were uplifted (Mountain and Prell, 1990). Figure 5.4 shows an example of seismic data from the Pakistan slope, which is dated by tying the reflectors to petroleum exploration wells on the Pakistan Shelf (Clift *et al.*, 2002b). This section shows some of the typical features of the submarine fan, most notably the prominent lens-shaped "channel–levee" complexes that dominate the upper part of the section, especially close to the shelf break. These represent constructional edifices built by successive turbidity flow events, flowing in the channel at the crest of the complex. The available age control shows that below the Lower Miocene there are no channel–levee complexes. The pattern of channel–levee construction testifies to a significant increase in the rate of sediment flux from the river after ∼24 Ma and peaking in the Middle Miocene (∼16 Ma; Figure 5.5).

Integration of many such dated lines has allowed a sediment budget to be generated for the Indus Fan, and this is shown in Figure 5.5. The budget shows a peak in sedimentation in the Middle Miocene and again in the Pleistocene, but with reduced rates in the Late Miocene and Pliocene. It is noteworthy that clay minerals in the sediments show a coherent development that may be interpreted in terms of changing monsoon weathering regime. The high values of kaolinite and especially smectite between 11 Ma and 4 Ma, albeit with a low around 8.5 Ma, suggest enhanced chemical weathering during the Late Miocene–Early Pliocene. Reduced physical erosion due to a more arid climate is consistent with the climate data from South Asia and the Indian Ocean discussed in Chapter 3, and would explain the reduced mass accumulation rates seen in the fan at that time.

At a first-order level the Indus sediment budget parallels the falling rates reconstructed by Burbank *et al.* (1993) for the Late Miocene (Figure 5.2) and suggests a common mechanism for erosion in the Indus and Bengal catchments. Because the drainage basins of the Indus and Bengal Fans are different in several aspects of their tectonic evolution this common pattern might reflect a dominant influence of the monsoon in controlling erosion. This hypothesis is further strengthened by Nd isotope data from sediments sampled from the Indus Fan. The Nd isotopes can be used as a first-order measure of the provenance of sediments because the isotopic values reflect the average value in the source regions and this ratio is not disturbed by erosion, weathering and transport (Figure 5.5; Clift and Blusztajn, 2005). In the Bengal Fan, a relatively steady Nd isotopic character since 17 Ma indicates constant sediment flux from the

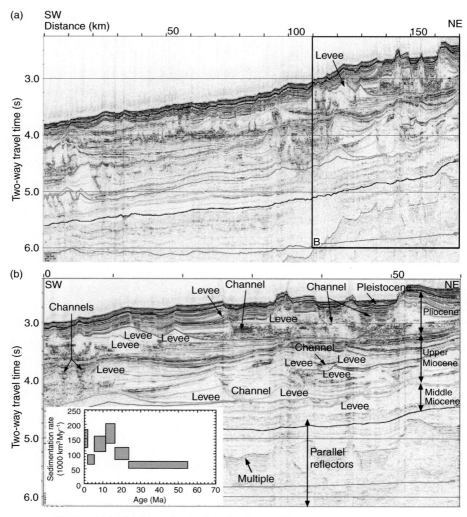

Figure 5.4 Multi-channel seismic-reflection profile SO122-26 from the upper Indus Fan offshore Pakistan showing the well developed channel–levee systems from Clift *et al.* (2002b). Note contrasting older conformable sediments of the underlying Paleogene, which show no obvious levee development and are interpreted as distal turbidites. (b) shows an enlarged up-slope portion of the line (black box in (a)) to demonstrate the geometry of the channel–levee systems. The insert shows the sedimentary budget calculated for the entire Indus Fan, based on this and other sections.

Greater Himalayas (France-Lanord *et al.*, 1993). The relatively consistent Nd isotopic character through much of the Neogene means that the source regions did not change their relative contributions substantially since ~30 Ma, a fact that is inconsistent with dramatic uplift of one range to feed the increased

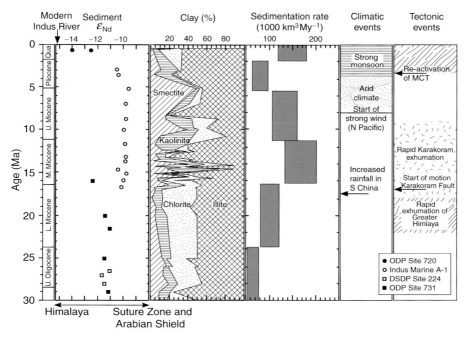

Figure 5.5 Diagram showing the evolving Nd isotope composition and sedimentation rates on the Indus Fan in relation to climatic and tectonic events known from onshore in Asia (from Clift and Blusztajn, 2005).

Middle Miocene sediment flux. Only in the Pleistocene does the Nd isotope character change greatly, a fact interpreted by Clift and Blusztajn (2005) to reflect large-scale capture of drainage from the Ganges into the Indus since 5 Ma, now represented by the major rivers of the Punjab (i.e., Sutlej, Jellum, Chenab and Ravi; Figure 5.6). However, the increased flux during the Pleistocene is only partly explicable by the addition of rivers to the trunk stream, and also requires faster erosion in the mountains driven by a stronger monsoon since 4 Ma. Similarly, the recognition of peak sedimentation in the Middle Miocene is consistent with climatic data that supports an early onset to the monsoon in the Early–Middle Miocene prior to the 8 Ma events.

5.5   Erosion in Indochina

If the monsoon is really controlling erosion of the Himalayas then its influence should also be seen in Indochina, where summer precipitation is now heaviest and presumably where the associated erosion would also be strong. Some indication of the influence of the monsoon has been provided by a study of where sediments in the modern Red and Mekong Rivers are eroded from.

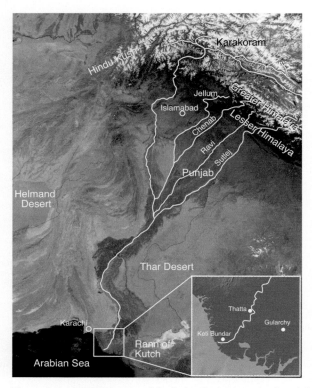

Figure 5.6 A MODIS image of the Indus drainage reproduced from NASA showing the major source regions and geographic features in the region. Drill sites are located close to the mouth of the modern Indus. Image reproduced with permission of NASA.

An array of Ar-Ar mica dating, U-Pb dating of zircon and fission track methods now suggests that crystalline rocks and especially inverted sedimentary basins on the flanks of the Tibetan Plateau are the most significant sediment sources (Clift *et al.*, 2006b). While these regions do experience a strong summer monsoon they are not the areas of the Red and Mekong catchments with the heaviest summer precipitation. This study concluded that active rock uplift, albeit in the presence of monsoon rains, was crucial to effective bedrock erosion.

Controls on erosion over longer timescales can be assessed by looking at the sedimentary record of flux into the ocean from the many rivers in East Asia. Unfortunately, the situation here is more complex than in the Indus, which is effectively the only major river in the western Himalayas. Figure 5.7 shows that the major drainage basins in East Asia, which mostly originate in eastern Tibet, spread out and discharge their sediment loads into the deltas of East Asia. It is noteworthy that many of the courses lie close together in eastern Tibet and Yunnan, just to the east of the eastern Himalayan syntaxis (Figure 5.8) and that

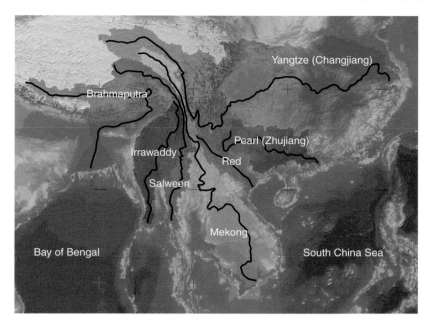

Figure 5.7 Shaded topographic relief map of East Asia showing the modern drainage basins of the largest rivers that bring material from eastern Tibet to the marginal seas. Note that the river courses all run close together in eastern Tibet, close to the eastern syntaxis of the Himalaya, making drainage capture from one system to another relatively simple. Figure modified from Clark *et al.* (2004). Reproduced with permission of the American Geophysical Union. See color plate section.

the pattern of drainages is not dendritic in the way that a river in equilibrium would be. Clark *et al.* (2004) interpreted this curious arrangement to reflect the north-eastward propagation of the Himalayan syntaxis and the general eastward growth of Tibet. Together these processes would have compressed the original drainage system, and provided the opportunity for one river to steal drainage from others through headwater capture.

Evidence for river capture in East Asia originally came from analysis of the nonsteady state drainage patterns seen in eastern Tibet and SW China (e.g., Brookfield, 1998; Clark *et al.*, 2004). This hypothesis is supported by the observation that there is more than twice as much sediment in the basins filled by the Red River (i.e., the Hanoi Basin and the Song Hong-Yinggehai Basin) than had been eroded in the modern drainage (Clift *et al.*, 2006a), implying that this river had lost drainage during the Cenozoic. When that capture occurred is only loosely constrained, although there does appear to be some consensus that this is driven by the surface uplift of the Tibetan Plateau. Clark *et al.* (2004) proposed that the original drainage was centered around an ancestral Red River from which progressive capture has removed most of the former headwaters. Analysis of the

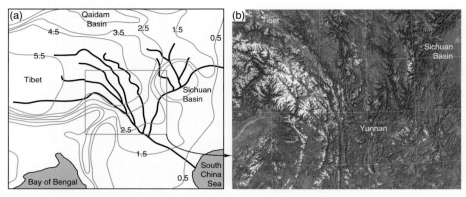

Figure 5.8 (a) Diagram redrawn from Clark *et al.* (2004) showing the configuration of the supposed pre-capture drainage of East Asia prior to major drainage capture around the eastern syntaxis of the Himalaya. The elevation contour is shown in kilometers. The gray box depicts the area shown in (b) as a satellite image from NASA Worldwind. Note the multiple parallel valleys incising the edge of the plateau and which are excavated by the heavy monsoonal rains that characterize the area.

sediments delivered by the Red River system offers the best chance to document changing erosion under the strong influence of the changing monsoon. Not only is this the "mother" drainage of East Asia, but also the region is well covered by seismic reflection surveys and industrial drill sites. Evolution in Nd isotope composition of sediments from the Red River delta now shows that while the sediments in the Eocene Red River were quite different from their modern counterparts there is no resolvable difference, at least in this isotope system by around 24 Ma (Clift *et al.*, 2006a). This implies major drainage capture prior to this time, presumably driven by the initial stages of Tibetan uplift. Mismatches between eroded and deposited volumes, however, indicate that drainage capture likely continued until Plio-Pleistocene times.

Models that propose either faster erosion driven by stronger monsoon rains or drainage capture make predictions about the composition and volume of sediment in any given river and thus the offshore sediment record can theoretically be used to reconstruct and date changes in erosion and climate. Figure 5.9 shows an example of a seismic section from south of Hainan island, showing the original faulted, rifted basement buried under thick sediment eroded both from Hainan and from eastern Tibet and delivered by the Red River. What is most striking is the development of huge southward-migrating deltaic foresets into the deep-water basin. Ties to petroleum wells show that these deltas are largely Plio-Pleistocene in age. Similar features are also seen further north in the Gulf of Tonkin, where the Middle Miocene is also very thick, similar to what was seen in the Arabian Sea. Figure 5.10 shows the Red River offshore region after monsoonal

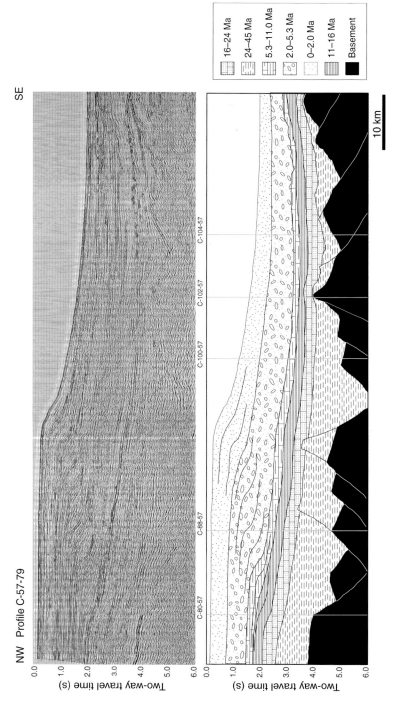

Figure 5.9 Multi-channel seismic profile C-79-57 released by BP Exploration and CNOOC from southern Hainan Island in the South China Sea, showing the original data above and interpretation below. This profile shows many of the features typical of the south Hainan margin and the Qiongdongnan Basin. The basement structure is typical tilted fault block style, while the sedimentary cover is dominated by thick, mostly Plio–Pleistocene foresets, migrating south into deep water. Reprinted with permission of the American Geophysical Union.

171

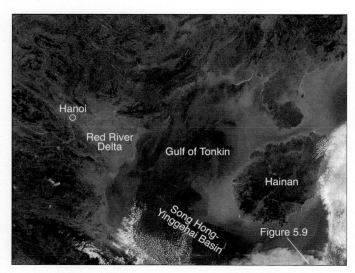

Figure 5.10 A gray scale copy of a MODIS true color image from NASA of the Gulf of Tonkin region showing the dominant character of the Red River delta in supplying sediment offshore. Recent rains have also triggered plumes of sediment to be eroded from Hainan itself, contributing to the seismic profile shown in Figure 5.8. Images reproduced with permission of NASA.

rains, especially on Hainan Island, which has been strongly affected by tectonic and volcanic activity in the last 1–2 My. Strong sediment plumes are derived from the Red River delta and from Hainan, both regions heavily influenced by summer monsoon rains. As a result of the recent tectonism Hainan is now a more important source of sediment to the South China Sea than it has been over the longer geological past.

In a regional synthesis Clift *et al.* (2004) used seismic data to derive sediment budgets from all the major drainage systems in East Asia (Figure 5.11). Because seismic data coverage is not uniform, the quality of the budgets is much better for the Mekong, Pearl and Red Rivers and less good in the Gulf of Thailand and East China Sea. However, even in these latter cases the long profiles available still allow a first-order constraint to be placed on the erosion rates for much of East Asia. The Gulf of Thailand is anomalous in this compilation, because although it is now fed by the Chao Phraya River it appears to have been mostly filled with sediment by around 11 Ma, after which time sediment flux was diverted to other basins. Because of the drainage capture issue it is most useful to examine the total erosional flux from East Asia if we are to understand the relationship between erosion and evolving monsoon strength. Summing all these sediment masses together, Clift *et al.* (2004) showed that erosion in East Asia increased gradually since ~45 Ma, rising sharply after ~33 Ma to reach a peak ~14 Ma

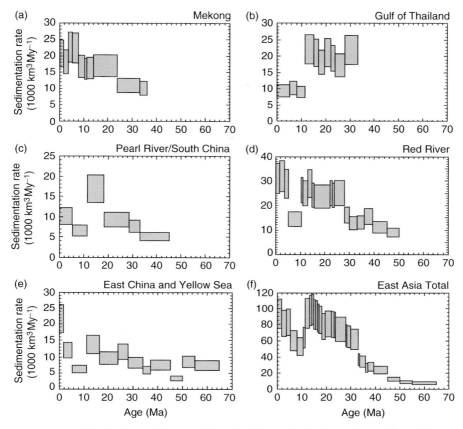

Figure 5.11 Sediment budgets for each of the major basin systems in East Asia, as reconstructed by Clift *et al.* (2004) (a) Mekong, (b) Gulf of Thailand, (c) South China margin, (d) Red River-Gulf of Tonkin and (e) Eastern China. (f) The proposed integrated sediment budget for all these basins, representing the net flux of material from Asia into the marginal seas. Reprinted with permission of the American Geophysical Union.

before decreasing to a low during the Late Miocene-Early Pliocene (Figure 5.11(f)). Erosion then rebounded to high values again in the Plio-Pleistocene.

Worldwide rapid erosion during the Plio-Pleistocene has been recognized by several authors (Zhang *et al.*, 2001; Métivier *et al.*, 1999) and has been linked to faster erosion in a variable glacial–interglacial climate (e.g., Molnar and England, 1990; Molnar, 2004). It is noteworthy that the modern clastic loads of the rivers in East Asia greatly exceed their average values during the Pleistocene (Clift, 2006). This means that if during the present Holocene period erosion rates are fast then they must have been slower in the past, presumably during glacial times, which are associated with weaker summer monsoons and stronger, drier winter monsoons. If glacial–interglacial cyclicity is the process driving enhanced Plio-Pleistocene erosion then another trigger must underlie the fast erosion in the Middle Miocene, because this period precedes Northern Hemispheric Glaciation at 2.7 Ma (Zachos *et al.*, 2001). In any case glaciation did not greatly affect many of the subtropical drainage basins in East Asia. Instead erosion rates at a continental scale may be more closely related to monsoon strength. New climatic data, mostly in the South China Sea, supports a wetter, more erosive climate in the Early–Middle Miocene, linked to earlier monsoon intensification (Clift *et al.*, 2002a; Jia *et al.*, 2003; Sun and Wang, 2005; Clift, 2006), which, when coupled with progressive plateau surface uplift, drove faster erosion.

5.6   Erosion in other regions

Further support for the monsoon being the primary driver of erosion in East Asia comes from noting the overall similarity between erosion records derived from that area with those from the Indus Fan (Figure 5.5). Even further afield erosion rates in the European Alps show a similar overall temporal pattern (Figure 5.12) that cannot be readily explained as reflecting synchronous tectonic activity. Most likely the similarity stems from changes in the global climate that in turn influenced monsoon strength. The Middle Miocene (∼16 Ma) represents a period of warm global conditions (Zachos *et al.*, 2001), which would favor stronger summer monsoons, while the Late Miocene–Pliocene is a time of cooler

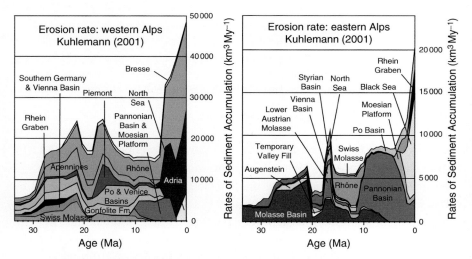

Figure 5.12 Masses of rock eroded from (a) the western and (b) the eastern Alps summed by considering the basins in which the eroded material was deposited; from Molnar (2004) and redrawn from Kuhlemann (2001) and Kuhlemann *et al.* (2001). Reprinted with permission of Elsevier B.V.

drier conditions as Antarctic glaciation intensified (Gupta *et al.*, 2004). With the onset of icehouse glacial–interglacial cycles, erosion strengthened again, although with most of the flux to the ocean being concentrated during the warmer, interglacial periods of high summer monsoon precipitation. At a first-order level there appears to be a correspondence between periods of strong summer monsoon intensity and enhanced continental erosion, that because of teleconnections in the global climate system are also reflected in changes of climate and erosion in other regions.

5.7   Monsoon rains in Oman

A dramatic example of the links between a strong monsoon and enhanced erosion is found in the interior bajada of Oman. The bajada is a huge array of alluvial fans, about an order of magnitude larger than those known from the present day (Figure 5.13). This bajada is no longer active, but represents

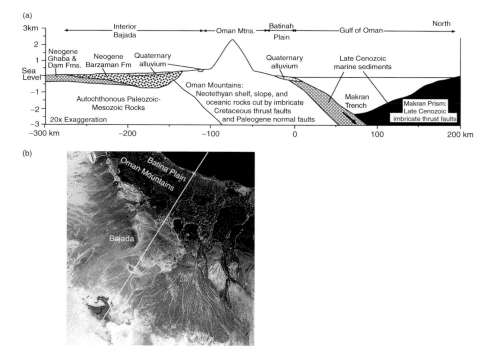

Figure 5.13 Cross-section from northern Oman to the Makran Prism showing four main geologic provinces, from Rodgers and Gunatilaka (2002). The cross-section is shown on a Landsat image. (a) Geologic cross-section showing schematic distribution of Late Cenozoic map units over older rocks. Data from Béchennec *et al.* (1993) and Calvache-Archila and Love (2001). Reprinted with permission of Elsevier B.V.

a period of fast erosion of the Oman Mountains during the Neogene. Rodgers and Gunatilaka (2002) proposed that aggradation of the bajada in Oman was largely facilitated by the wet–humid monsoonal climate that characterized the latest Miocene, Pliocene and Early Pleistocene in Arabia. In contrast, the dominant arid or semiarid phases of the Late Pleistocene led to fan erosion and deflation. The paleo-climate of the Arabian interior is not so well known as the detailed Holocene records based on cave records discussed in Chapter 4, but has been reconstructed based on fan morphology (Maizels and McBean, 1990), isotopic data from carbonate cements in the bajada (Burns and Matter, 1995), together with isotopic and pollen data from oceanic sediments along Oman's south-eastern margin, which track the flora of Arabia and the strength of the monsoon winds (Nitsuma *et al.*, 1991; Van Campo, 1991; Gupta *et al.*, 2003). These data imply that monsoon rainfall frequency and intensity had a much more northerly latitudinal range in Arabia in the Late Miocene–Early Pleistocene than in the present. With a southerly shift of the monsoon track away from the Arabian Peninsula, probably in the Late Pleistocene, increasing aridity set in, so reducing erosion. In Oman, as in Asia, there appears to be a close correlation between monsoon rains and erosion.

5.8   Changes in monsoon-driven erosion on orbital timescales

The impact of monsoon strengthening on continental weathering may differ in different regions, depending on the nature of the monsoon in those regions and on the topography of the drainage. More rain might result in faster erosion and enhanced physical weathering in mountainous regions, yet in low lands more rain could drive more intense chemical weathering. Here we examine the impact that millennial-scale changes in climate have had on weathering and erosion in South and East Asia.

5.8.1   *Ganges–Brahmaputra Delta*

Enhanced erosion linked to a strong monsoon is clearly shown in the Holocene stratigraphy of the Ganges–Brahmaputra Delta. High-resolution seismic imaging and coring of channel–levees on the upper Bengal Fan allow sediment accumulation rates on these features to be traced since the Last Glacial Maximum. Figure 5.14 shows an example of an active channel–levee with the changes in sedimentation rate calculated from dating of cores (Weber *et al.*, 1997). It is noteworthy that sedimentation increased rapidly after 13 ka, after the Younger Dryas and as the climate warmed and wettened onshore. The sharp fall in sedimentation rates around 9.5 ka only reflects the fact that the rising sea level was providing accommodation space onshore for the sediment flux,

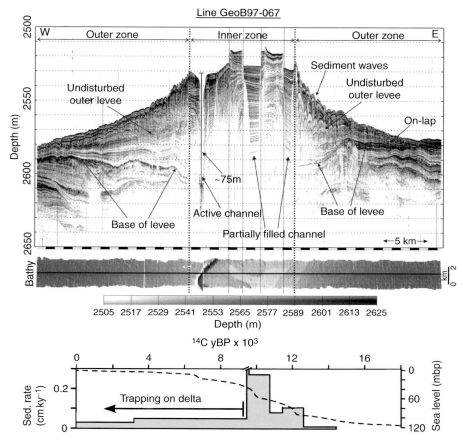

Figure 5.14 Seismic reflection (3.5 kHz Parasound) records and sedimentation history for the active channel–levee system of the upper Bengal Fan. Accretion of the channel– levee fill begins in the early postglacial and greatly increases toward the hypsithermal with increased river discharge. The precipitous drop in sedimentation after 9.5 ky results from a shift of the depocenter onto the Bengal margin owing to rising sea level (modified from Weber *et al.*, 1997).

not that erosion was getting slower. Fortunately, coring onshore by Goodbred and Kuehl (1999, 2000) was able to sample and then date the Holocene delta sediments that correspond to the deep-water fan deposits (Figure 5.14). Accumulation rates in the delta increased sharply from 9 to 7 ka, reflecting, first, the shift in depocenter as the sea level changed and, second, the fall due to moderate drying of the climate after an 8 ka maximum in monsoon precipitation (Herzschuh, 2006). Although rainfall in the Ganges–Brahmaputra basin remains high to the present day it is less than in the mid-Holocene. A positive relationship

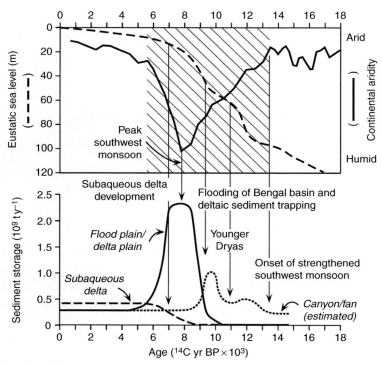

Figure 5.15 Plots of eustatic sea level (after Bard *et al.*, 1996; Fairbanks, 1989), South Asia aridity index (from Prins and Postma, 2000), and Ganges–Brahmaputra sediment storage (t is metric tons; values based on Goodbred and Kuehl, 1999, 2000; Weber *et al.*, 1997). Hatched area shows period of high regional insolation (Colin *et al.*, 1999). Vertical arrows and annotations highlight correlations between climate, sea level, and events in the Ganges–Brahmaputra fluvial delta system. Note that timescale is in radiocarbon, not calendar, years. Reprinted with permission from the Geological Society of America.

between continental erosion rates and summer monsoon intensity is clearly shown in this system, consistent with the record over longer time spans during the Neogene (Figure 5.15).

Coring in the Andaman Sea and Bay of Bengal provides a weathering record for the Ganges–Brahmaputra and Irrawaddy Rivers over the last several marine isotopic stages, allowing us to document the weathering response to change in monsoon strength over several glacial cycles. Colin *et al.* (1999) analyzed clay minerals from these cores, which when coupled with oxygen isotopic data from the associated foraminifers can be correlated to global marine isotope stages, and thus provide a reliable climate record. Figure 5.16 shows that a clear pattern emerges, with more smectite (and kaolinite) being produced in both river systems during interglacial times, compared with more illite and chlorite during glacial times.

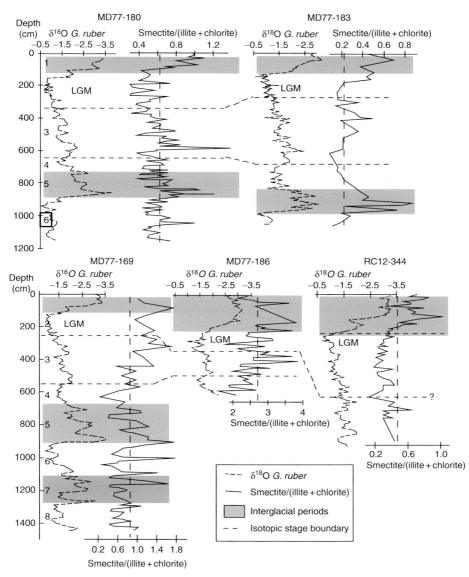

Figure 5.16 The δ^{18}O of *G. ruber* and smectite/(illite + chlorite) ratio versus depth (cm) for cores from Colin *et al.* (1999): RC12-344 and MD77-169 (Andaman Sea), MD77-180, MD77-186 and MD77-183 (Bay of Bengal). Isotopic stages are also reported. Reprinted with permission of Elsevier B.V.

These data were interpreted to indicate that the wet summer monsoons during interglacial times favored stronger chemical weathering and thus smectite and kaolinite production in vegetated soils, mostly on alluvial plains. In contrast, their production was reduced in the drier glacial times. The Irrawaddy basin responds to monsoon intensification by increased chemical weathering because the sediment sources are not very mountainous and do not expose many high-grade metamorphic rocks. In the tropical climate the dominant response to summer monsoon strengthening is not greater physical erosion. However, it is also noteworthy that the Bengal Fan cores also show the shift to more chemically weathered smectite during the interglacial times, at least during the initial deglaciation period. The relative contribution of smectite falls during interglacials after an initial maximum, perhaps reflecting increasing physical weathering as a strong summer monsoon continues.

The source and weathering intensity of the sediments deposited in the Andaman Sea and Bay of Bengal during the Last Glacial Maximum can also be assessed using isotopic methods. These sediments are characterized by a significant increase in $^{87}Sr/^{86}Sr$ with little change in ε_{Nd} (0) (Colin et al., 1999). This Sr isotopic composition change is associated with a decrease in the smectite/(illite + chlorite) ratio, consistent with a decrease in chemical weathering intensity. These variations can be attributed to a $^{87}Sr/^{86}Sr$ change of the detrital material carried by the Indo–Burman rivers to the Bay of Bengal and the Andaman Sea. Radiogenic $^{87}Sr^{+2}$ ions created from $^{87}Rb^+$ are located in a different lattice configuration than the nonradiogenic Sr ions incorporated during the primary crystallization of the mineral, which is referred to as 'initial Sr'. As a consequence, Rb-rich minerals will have a tendency to release ^{87}Sr preferentially during weathering, thus shifting the $^{87}Sr/^{86}Sr$ value during deglaciation (Colin et al., 1999).

5.8.2   Indus Delta

Drilling at a series of sites close to the modern Indus delta mouth now allows the erosional response of the Indus drainage to changing monsoon intensity since the Last Glacial Maximum to be assessed (Figure 5.6). This work shows two coarsening-upward cycles, separated by a transgressive mud deposited after ~8 ka (Giosan et al., 2006). Facies analysis and sequence stratigraphy methods indicate that the delta prograded in two phases, at 14–8 ka and 8 ka to the present. Although the shoreline retreated ~30 km northwards at 8 ka, the delta was prograding southwards during the 12–8 ka period when rates of eustatic sea level rise were greatest (Camoin et al., 2004; Siddall et al., 2003). This behavior contrasts with most deltas at this time and suggests that sediment flux to the Indus delta was able to keep pace with the rising sea level, similar to the situation in the Ganges–Brahmaputra delta (Goodbred and Kuehl, 2000).

Sediment samples from the boreholes make it possible to assess the changing erosional flux of the river as the monsoon intensified. X-ray diffraction analysis of clay minerals in the sediment shows no change in the composition since the Last Glacial Maximum, being dominated by chlorite and illite, indicative of

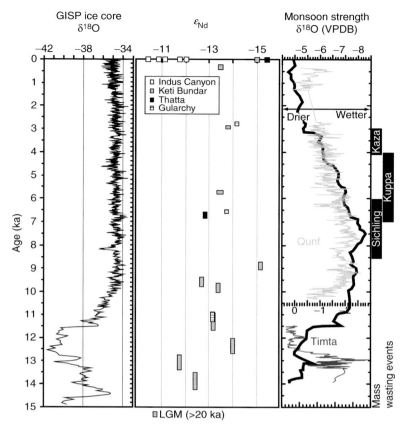

Figure 5.17 Diagram showing the variability in Nd isotopes in the Indus delta (see Figure 5.6 for locations) since the Last Glacial Maximum compared with the $\delta^{18}O$ isotope record from the GISP2 ice core of Stuiver and Grootes (2000), the intensity of the SW monsoon traced by speleothem records from Qunf and Timta Cave speleothems (Fleitmann *et al.*, 2003; Sinha *et al.*, 2005) in Oman and by pollen (Herzschuh, 2006) from across Asia (black line), as well as western Himalayan landslides (Bookhagen *et al.*, 2005a). The diagram shows the relationship of erosion to climate change and monsoon strength during the Holocene. Modified from Clift *et al.* (2008). Reproduced with permission of the Geological Society of America.

a dominant physical weathering regime in the sources. However, Nd isotope analysis of the sediments in the Indus delta shows how the relative proportion of sources feeding the Indus have changed since 14 ka (Figure 5.17). What is surprising is that there is almost as much variability since 14 ka as seen in the entire 30 Ma reconstruction shown in Figure 5.5. This variation does not show any coherent differences in provenance between sediment of different grain size and points to large-scale changes in the sediment sources since the Last Glacial Maximum. Although regional drainage capture has been invoked to explain the isotopic changes seen in Figure 5.5 (Clift and Blusztajn, 2005), this explanation is not practical for explaining every such change in the history of the Indus. There is also no geomorphologic evidence to suggest major drainage capture since 14 ka (e.g., Ghose *et al.*, 1979).

The Nd isotope data, supported by single grain U-Pb zircon dating, indicate that since 14 ka the Indus River started to receive much greater volumes of detritus from Lesser Himalayan sources relative to the Karakoram (Clift *et al.*, 2008), which is the dominant source during glacial times (Clift *et al.*, 2002c). The greatest change in provenance occurred around 12–13 ka, around the end of the Younger Dryas, although peak ε_{Nd} values were dated around 9 ka, the time of maximum summer monsoon intensity according to Prins and Postma (2000). If drainage capture is not the primary driver of this change then this must reflect changes in the efficiency of erosion in the sources. The Himalayan front is the location where most monsoon precipitation now falls (Bookhagen and Burbank, 2006), and it follows that it is in that area that the greatest changes in erosion would occur as the summer monsoon strengthened. In contrast, the Karakoram lie mostly in the rain shadow of the Himalayas and do not receive much monsoon precipitation. Instead hydrogen and oxygen isotope analysis indicates that they receive most of their moisture from the west, from Mediterranean sources (Karim and Veizer, 2002). In this region, erosion is largely accomplished by glacial scouring. As a result of summer monsoon strengthening the whole provenance of the river has changed dramatically since 14 ka.

5.8.3   *South China Sea*

Similar weathering patterns are also revealed in the South China Sea over millennial timescales during the Pleistocene. Liu *et al.* (2003) used mineral assemblages from ODP Site 1146 in the northern South China Sea to track strong glacial–interglacial cyclicity. Figure 5.18 shows that on the passive margin of southern China high concentrations of illite, chlorite and kaolinite are found during glacial periods, while high smectite and mixed-layer clay are associated with interglacials. The situation is apparently more complex in the South China Sea than the South Asian deltas because Liu *et al.* (2003) proposed that most of the smectite on the South China margin is brought by currents as products from the chemical weathering of volcanic rock sources in Luzon, located to the south-east. In contrast, along-shore transport from Taiwan and the Yangtze River is considered to be the dominant provider of illite and chlorite. Kaolinite is mostly supplied directly from the neighboring Pearl River. Eolian transport has been proposed as a further source of material, with quartz and feldspar derived largely from central Asia via winter monsoon winds (Tamburini *et al.*, 2003).

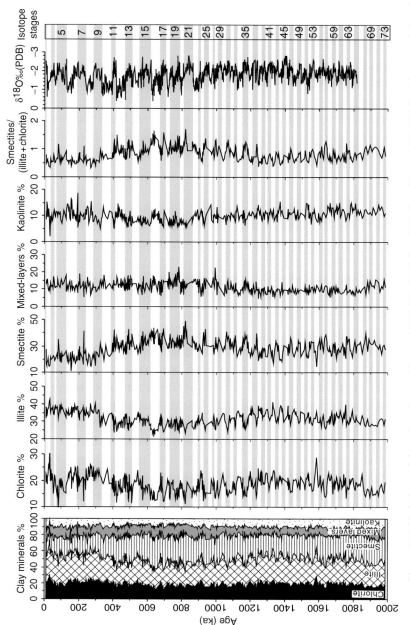

Figure 5.18 Variations of clay mineral assemblages at ODP Site 1146 since 2 Ma from Liu et al. (2003). Mixed layers mainly refer to random smectite–illite and smectite–chlorite mixed layers. Isotope stages were obtained by visually correlating $\delta^{18}O$ (Clemens and Prell, 2003) (stages 1–65) and smectites/(illite + chlorite) ratio (stages 66–73) to $\delta^{18}O$ records of ODP Site 677 (Shackleton et al., 1990). Reprinted with permission of Elsevier B.V.

In this area, changes in the proportion of different clay minerals only partially reflect changes in the weathering regimes in the source regions, but are principally driven by changes in the relative strength of the transport processes. During interglacial periods, summer enhanced monsoon (south-westerly) currents would transport more smectite and mixed-layer clays to ODP Site 1146, whereas during glacial periods, enhanced winter monsoon (northerly) currents transport more illite and chlorite from Taiwan and the Yangtze River. Clift *et al.* (2002a) had argued for a simpler provenance being dominated by the flux of material from the Pearl River with changes in mineralogy being triggered by weathering changes, similar to that proposed by Colin *et al.* (1999). Whatever the cause it is clear that the mineralogy of sediments on the Chinese margin is strongly linked at orbital timescales to the relative strengths of the summer and winter monsoons, and that the weathering and erosion of Asia is primarily linked at these timescales to monsoon processes.

5.9   Tectonic impact of monsoon strengthening

If strengthening of the Asian monsoon system is largely related to topographic uplift of Tibet, then it is also worth considering whether there is any feedback on solid Earth tectonics resulting from monsoon intensification. Above we have described how the intensity of the monsoon has been a dominant control on the Cenozoic erosion of Asia, both in East Asia, where the river systems incise the flank of the Tibetan Plateau, and in South Asia, where the tectonically driven rock uplift of the Himalayas may also be a key factor controlling erosion. What is clear is that at a regional scale there is a close correspondence between topography, active faulting and precipitation. Figure 5.19 shows the variability in precipitation across central Tibet, demonstrating how the heaviest rains occur at the Himalayan front, but just south from the very highest topography, which is largely free from precipitation. There is also a peak in precipitation along the northern edge of the plateau, although it is not so intense because the monsoon rains from the Indian Ocean are much greater than those that reach the interior of Asia. As a result of this pattern most of Tibet can be seen to lie in a deep rain shadow, with minimal precipitation over most of the plateau. It is also noteworthy that the peaks in precipitation correspond to ranges of mountains that exceed the average altitude for the bulk of the plateau.

The key question is whether these bounding ranges exist because of tectonic processes, so that the rain then falling at the topographic boundary can be thought of as simply orographic precipitation, or whether the rains themselves and their enhanced erosion causes the uplift of the mountains in the first place. Certainly the concept that climate might cause mountain uplift is not new.

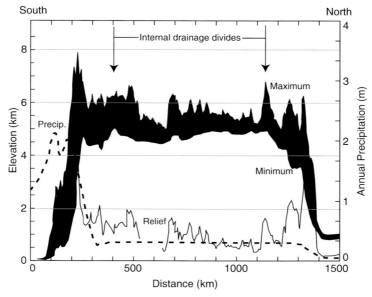

Figure 5.19 Diagram from Fielding (1996) showing the relationship between topography and precipitation across the Tibetan Plateau, with heaviest rainfall along the southern Himalayan front and to a lesser extent on the northern Kun Lun–Altun Tag margin. Note that central Tibet has a very arid climate. Shaded regions show the range of maximum and minimum altitudes. The dashed line indicates annual precipitation. Reprinted with permission of Elsevier B.V.

The apparent global reactivation of mountain chains in the Pleistocene may be linked to a more erosive climate since that time, specifically to the onset of Northern Hemispheric Glaciation (Molnar and England, 1990; Zhang *et al.*, 2001). The basic concept is that stronger erosion in valley floors removes material, unloads the crust and allows the adjacent peaks to rise in order to reach isostatic equilibrium. This model has been applied over small regions to explain the dramatic metamorphism and rapid uplift of the Nanga Parbat massif in the Pakistan Himalayas (Koons *et al.*, 2002, Zeitler *et al.*, 2001). In that case the uplift is believed to be caused by the efficient removal of rock by the Indus River, though this could be accomplished on a regional scale by smaller rivers. Whether this logic is universally applicable is not clear, as Whipple *et al.* (1999) demonstrated that in tectonically active mountain ranges, geomorphic constraints allow only a relatively small increase in topographic relief in response to climate change. Thus, although climate change may cause significant increase in denudation rates, potentially establishing an important feedback between surficial and crustal processes, neither fluvial nor glacial erosion is likely to induce significant peak uplift.

Links between monsoon rains and tectonic deformation have recently focussed on models that invoke a 'stream-power' rule, in which either increased discharge or steeper channel slopes cause higher erosion rates in the presence of strong monsoon rains. Spatial variations in precipitation and slopes are therefore predicted to correlate with changes in both erosion rates and crustal strain. In practice this means that patterns of rainfall may control the distribution and intensity of tectonic faulting as precipitation removes rock mass at the mountain front and allows deeper buried units to uplift to the surface in those areas (Wobus et al., 2005). A study by Hodges et al. (2004) from the Annapurna region of the Nepalese Himalayas shows the proposed relationships clearly (Figure 5.20). In that region the zone of recent faulting is coincident with an abrupt change in the gradient of the Marsyandi River and its tributaries, which is thought to mark the north to south transition from a region of rapid uplift in the Greater Himalayan ranges to a region of slower uplift in the Lesser Himalayas.

If monsoon-driven erosion is responsible for the faulting and formation of a topographic break then uplift of the Greater Himalayas during the Quaternary is not entirely due to passive uplift over a deeply buried ramp in the Himalayan sole thrust, as has been typically proposed. Instead Himalayan uplift must partially reflect active thrusting at the topographic front. The coincidence of active thrusting with intense monsoon precipitation suggested to Hodges et al. (2004) that a positive feedback relationship is operating between focussed monsoon-driven erosion and deformation at the front of the Himalayan ranges.

An alternative view of how monsoon rainfall and Himalayan tectonics interact was taken by Burbank et al. (2003) who attempted to quantify coupling between the two by comparing meteorological observations from across the same region of the Marsyandi River catchment in Nepal with estimates of erosion rates at timescales greater than 100 ky (Figure 5.21). Erosion rates were derived from low-temperature thermochronometry, mostly apatite fission track, that is sensitive to cooling from temperatures of 60–110 °C (equivalent to 2–5 km burial; Green et al., 1989). Burbank et al. (2003) concluded that across a zone of about 20 km width, including the highest ranges of the Himalayas, significant spatial variations in erosion rates are not detectable. The conclusion even included zones with huge (fivefold) differences in monsoon precipitation. Furthermore, decreased rainfall did not appear to be balanced by steeper channels. Instead, additional factors must have influenced river incision rates. Channel width and sediment concentrations were inferred to compensate for decreasing precipitation. Burbank et al. (2003) concluded that because they had documented spatially constant erosion that did not appear to respond directly to variations in monsoon precipitation then erosion in the Nepalese Himalayas must be a response to solid Earth tectonic transport of Greater Himalayan rock above a crustal ramp.

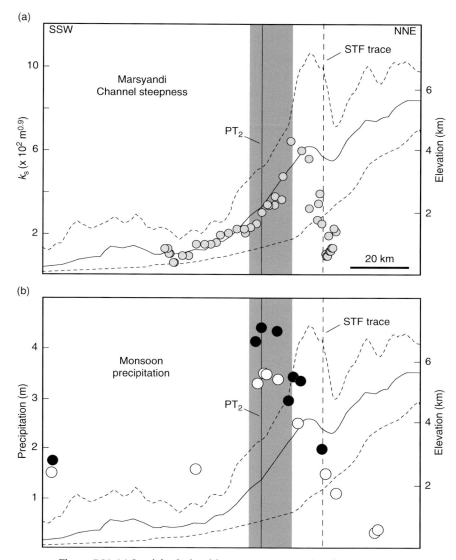

Figure 5.20 (a) Spatial relationships among topography, river steepness and deformational features in the Annapurna Range of Nepal from Hodges *et al.* (2004). The thin solid line indicates mean elevations along a 50 km wide, SSW–NNE transect orthogonal to the range front, with dashed lines indicating the minimum–maximum envelope; the scale is shown on the right. Shaded circles indicate values of k_s, the steepness index, (the scale is k_s shown on the left) based on slope–drainage area data for the Marsyandi trunk stream at 5 km intervals along its course. Because the river does not run SSW–NNE along its entire course, each point has been projected to the line of section. The shaded region indicates the boundaries of the Quaternary deformation. The thick vertical line indicates the position of the Physiographic

While these two studies of effectively the same region seem to come to quite different conclusions about the role of the monsoon in driving Himalayan erosion and faulting, in detail the discrepancy may relate to timescales. The thermochronologic fission track data do show that there is no one-to-one spatial correlation between modern heavy monsoon rainfall and long-term exhumation, but this does not mean that there can be no link between monsoon strength and tectonic activity. It is noteworthy that at a scale of 50–100 km there is good correspondence between the highest rainfall and most active exhumation (Figures 5.17 and 5.18). Wobus *et al.* (2005) used *in situ* cosmogenic ^{10}Be data to demonstrate a fourfold increase in millennial timescale erosion rates, over a distance of less than 2 km in central Nepal. By doing this these workers were able to delineate for the first time an active thrust fault nearly 100 km north of the surface expression of the Main Himalayan Thrust, which is presumed to account for most of the recent active compression. Crucially, this newly recognized thrust lies in a zone of very high monsoonal precipitation. Wobus *et al.* (2005) argue that these new data prove the idea that rock uplift gradients across the Himalaya in central Nepal are connected with monsoon precipitation, at least over short periods of geologic time, and note that this pattern of recent exhumation is not consistent with passive transport over a ramp in the Main Himalayan Thrust.

Related work by Thiede *et al.* (2004) in the Sutlej drainage of the western Himalayas broadly supports these conclusions of major climatic control on orogenic structure. Figure 5.22 shows a cross-section through the Himalayas in the vicinity of the Sutlej Valley and compares topography, annual precipitation (mostly summer monsoon-related) and tectonic structure. As noted above, the precipitation tends to fall at the break in slope of the topography, which in the Sutlej region is in two locations, at the transition between the foreland basin, and between the Lesser and Greater Himalayas. The correspondence between precipitation and major faults is striking, with the Main Boundary Thrust (MBT) and an outlier of the Main Central Thrust (MCT) located over the southern

Caption for Figure 5.20 (*cont.*)

Transition 2 (PT2), dividing the Greater and Lesser Himalayas, as defined by stream profile analysis. The dashed line marks the approximate trace of the basal detachment of the South Tibetan (STF) fault system in the area. (b) Monsoon rainfall distribution across the Annapurna Range, projected to the same line of section shown in (a). Circles indicate average annual monsoon precipitation, as measured at meteorological stations on ridges (filled) and in valleys (open), projected onto the line of section; the scale is shown on the left. Reprinted with permission of Elsevier B.V.

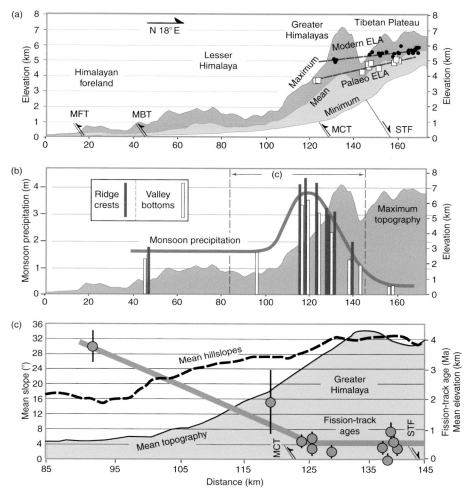

Figure 5.21 Monsoon precipitation, apatite fission-track ages, glacial equilibrium-line altitude (ELA) and topographic characteristics of the Marsyandi drainage from Burbank *et al.* (2003). MFT, Main Frontal Thrust. (a) Maximum, minimum and mean elevation within a 50 km wide swath oriented parallel to the strike of the range in the study area. Major faults (STF, MCT, MBT and MFT) are shown. Modern (black circles) and glacial-era (white squares) equilibrium line altitudes define northward-rising gradients with a steeper gradient during glacial times. (b) Monsoon precipitation projected against maximum topography through the study area. Monsoon rainfall is ~10 times greater on the southern Himalayan flank than north of the Annapurna crest, as also noted by Hodges *et al.* (2004). The spatial extent of (c) is indicated. (c) Apatite fission-track ages plotted against mean topography. Between the MCT I and the STF system, the apatite fission-track ages are consistently very young (<1 Ma). No apparent gradient in ages exists across the Greater Himalayas. Mean hillslope angles systematically increase northward across the zone of uniform cooling ages. Reproduced with permission of Macmillan Magazines Ltd.

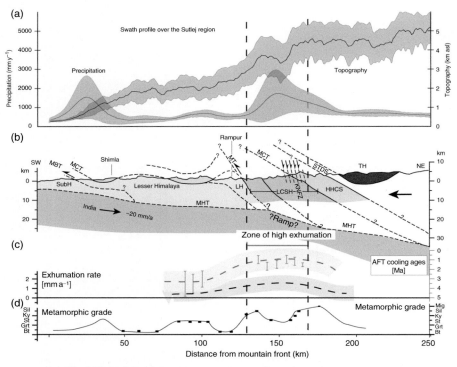

Figure 5.22 (a) Compiled data from the study of Thiede *et al.* (2004) illustrate the coupling between surface processes and tectonism in the Sutlej Region in the western Indian Himalayas. Topographic and precipitation-distribution swath profiles for the Sutlej area are oriented perpendicular to the southern Himalayan front. Swath profiles centered along the river are 250 km long and 100 km wide; thick lines indicate mean values, shaded areas denote maxima and minima. Topography based on (USGS), mean annual precipitation is derived from SSM/I passive microwave satellite data. Note that the distribution of orographic precipitation is focussed between elevations of ~2000 and ~3500 m in a ~50–70-km-wide zone. (b) The simplified geologic cross-section parallel to the swath profiles shows that rocks of the Lesser Himalaya Crystalline Series in the footwall of the Main Central Thrust are exhumed in this sector. (c) Parallel to the geologic cross-section are apatite fission track cooling ages along the Sutlej River shown (upper) and estimated exhumation rates (lower); dashed lines indicate mean values, shaded areas denote ±2σ. In panel (d) the metamorphic grade of the rock units along the profile is plotted (modified after Vannay *et al.* (1999)). Based on fission track cooling ages (c), the coincidence between rapid erosion and exhumation is focussed in a 50–70 km wide sector of the Himalayan deformation belt, rather than encompassing the entire orogen. We assume that the enhanced and focussed orographic precipitation (a) has localized rapid erosion and exhumation over geologic time. To accommodate this concentrated loss of material, deeper high-grade metamorphic rocks are exhumed by motion along a back-stepping thrust to the south (MT) and normal fault zone (KNFZ) to the north

precipitation zone, while the northern zone is located above another splay of the MCT, labeled MT on Figure 5.20. This northern MT fault allows higher-grade rocks to preferentially come to the surface in this area.

The observation that fission track cooling ages are uniformly younger in the zone of high precipitation is consistent with the hypothesis that it is the summer monsoon rains removing material from the Greater Himalayas that allows active exhumation to continue (Thiede *et al.*, 2004). This conclusion was further reinforced by a study of the apatite fission track and muscovite Ar-Ar cooling ages in the Greater Himalayas around Annapurna in central Nepal. These data showed that there was a fivefold increase in apparent erosion rate between 2.5 and 0.9 Ma at a time when the summer monsoon greatly strengthened (Huntington *et al.*, 2006). Because there is no evidence for major changes in the tectonics of the Himalayas at that time climatic (i.e., monsoonal) processes are invoked as being the primary controls on mountain exhumation.

5.9.1   *Erosion on millennial timescales*

As well as over longer periods of geologic time it is noteworthy that monsoon strength on millennial timescale intervals can be linked to mass wasting along the Himalayan front. It is the loss of material from over deep buried rocks that allows them to be uplifted to the surface in the Greater Himalayas. While some of this exhumation has been achieved by tectonic unroofing along detach-ment structures (e.g., Herren, 1987; Burchfiel *et al.*, 1992; Searle and Godin, 2003) this can also be achieved by erosional unroofing and unloading at the range front. Figure 5.23 shows the relationship between major landsliding events in the western Himalayas and the intensity of the monsoon over the last 40 ky, as recorded by a series of proxies discussed in Chapter 4 from Arabia, western India and southern China. It is apparent from this study that major landsliding events do not occur randomly, but are associated in time with periods of increased regional humidity linked to a stronger summer monsoon (Bookhagen *et al.*, 2005a). Such a correspondence should not be unexpected because mass wasting and landslides are closely correlated to precipitation intensity in active moun-tain ranges, such as Taiwan (e.g., Dadson *et al.*, 2003). Water-logged soils on steep

Caption for Figure 5.22 (*cont.*)

(b) that is localized by climatically controlled erosional processes. TH, Tethyan Himalayas; HHCS, High Himalayan Crystalline Sequence; LHCS, Lesser Himalayan Crystalline Sequence; LH, Lesser Himalayas; SubH, Sub Himalayas; STFS, Southern Tibetan Fault System; MCT, Main Central Thrust; KNFZ, Karcham Normal Fault Zone; MT, Munsiari Thrust; MBT, Main Boundary Thrust; MHT, Main Himalayan Thrust. Reprinted with permission of Elsevier B.V.

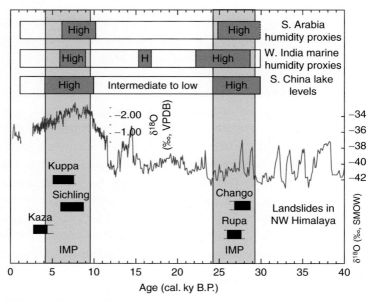

Figure 5.23 Diagram from Bookhagen *et al.* (2005a) showing the close relationship between mass wasting and landsliding events and the intensity of the Asian monsoon since 40 ka. Late Pleistocene and middle Holocene intensified monsoon phases (IMPs). Information on $\delta^{18}O$ measurements (VPDB – Vienna Pee Dee Belemnite; SMOW – standard mean ocean water) is merged from GISP2 (Greenland Ice Sheet Project, 1997) and high-resolution Holocene speleothem data from the southern Arabian Peninsula (Fleitmann *et al.*, 2003). Speleothem data indicate strengthened south-west monsoon during the late to middle Holocene. Multiple past humid phases in the desert of southern Arabia (S. Arabia humidity proxies) show two distinctive wet intervals (Bray and Stokes, 2004). Humidity proxies from the north and east Arabian Sea (W. India marine humidity proxies) show intensified summer monsoons (Prins and Postma, 2000; Thamban *et al.*, 2001). Independently derived chronologies of humid phases in adjacent areas (S. China lake levels) from lake highstands and pollen records emphasize the regional importance of these humid intervals (e.g., Fang, 1991; Gasse *et al.*, 1991). Black boxes mark the existence of landslide-dammed lakes in the greater Sutlej Valley region; black lines outside boxes signify age uncertainties. Reprinted with permission from the Geological Society of America.

mountain slopes are more likely to undergo mass wasting than their dry equivalents, while rivers swollen by summer rains erode steep gradients and undercut banks, further encouraging mass wasting. The enhanced flux of sediment to both the Indus and Bengal deltas, especially during the Early Holocene when the summer monsoon was especially strong, indicates that mountain erosion is closely coupled to monsoon strength over millennial timescales (Goodbred and Kuehl, 2000). The fact that the composition of the sediment in the Indus changes in phase with the monsoon (Figure 5.17) further eliminates the possibility that the rains are merely washing away material eroded by glaciers prior to intensification (Clift *et al.*, 2008). Changes in monsoon strength appear to result in effectively instantaneous erosional responses in the mountains.

5.10   Climatic control over Himalaya exhumation

If summer monsoon precipitation is a major control on the topography and tectonics of the Himalayas in the recent geologic past, is it worth asking how far back in time this connection might extend? The age of onset of rapid exhumation of the Greater Himalayas is not firmly constrained because dating of the rocks now at the surface in the high crystalline ranges only reveals when those particular rocks began their ascent and cooling, but this does not preclude earlier uplift. Earlier exhumed Greater Himalayan rocks would have been removed by erosion. Radiometric dating of the granites and metamorphic rocks that make up the Greater Himalayas indicate that rapid cooling started around 23 Ma (e.g., Hodges and Silverberg, 1988; Treloar *et al.*, 1989; Guillot *et al.*, 1994; Hodges *et al.*, 1996; Walker *et al.*, 2001; Godin *et al.*, 2006; Figure 5.24). Dating of rocks around the Main Central Thrust in the eastern Himalayas indicates a start of motion on that fault close to 22–23 Ma, effectively synchronous with the South Tibetan Detachment, but the record of earlier exhumation must lie in the erosional products. Unfortunately, there is no sediment dating between 35 and 24 Ma preserved in the Himalayan foreland basin owing to a major unconformity (Najman *et al.*, 2001; Najman, 2006). While these materials presumably lie offshore in the Indus and Bengal Fans they have not yet been sampled by drilling. Thus the precise age of the start of Greater Himalayan exhumation is not well defined, although it is presumed to have occurred just before 23 Ma.

The sedimentary rocks in the Indian foreland basin directly overlying the unconformity (known as the Dagshai Formation) are roughly dated as being younger than 30 Ma in western India (Najman, 2006). The Dharmsala Formation, also exposed in India, is constrained as being older than 21 Ma and is a correlative of the Dagshai Formation. Equivalent pre-Siwalik sedimentary rocks in Nepal are better dated, with magnetostratigraphic methods indicating a range of 16–21 Ma for the basal Dumri Formation (DeCelles *et al.*, 2001). Najman (2006) noted that the Dagshai Formation contains no metamorphic garnet grains, which are common in Greater Himalayan rocks, but that this mineral is found in the overlying sediments, dated as younger than 22 Ma. This change in mineralogy of the eroded sediments indicates that the depth of exhumation in the source regions increased during the transition upward from Dagshai and Dumri to the Kasauli Formation, i.e., that the Greater Himalayan were first exposed

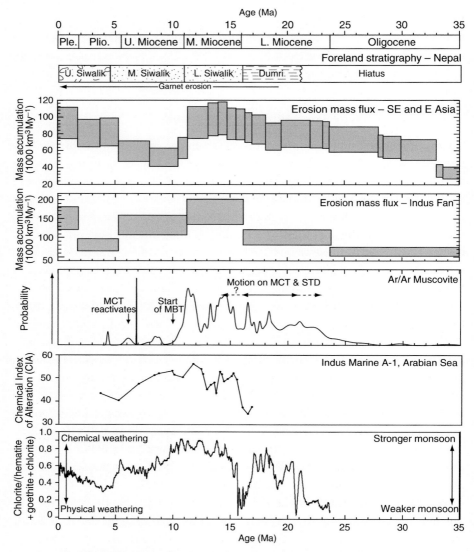

Figure 5.24 Diagram showing correlations between monsoon climate, clastic sedimentation rates (Clift, 2006), erosional history and the tectonic evolution of the Greater Himalayas since 35 Ma. Temporal evolution of clay mineralogy at ODP Site 1148 in the South China Sea is derived from unmixing of diffuse reflectance spectrophotometry data. The Chemical Index of Alteration (CIA) is calculated from the bulk chemical composition of cutting samples from Well Indus Marine A-1 on the Pakistan shelf using the method of Nesbitt and Young (1982). Foreland stratigraphies and age dates are simplified from Najman (2006). Muscovite Ar-Ar cooling ages are compiled from the literature. MCT = Main Central Thrust, STD = South Tibetan Detachment, MBT = Main Boundary Thrust.

fully around 21 Ma, but were likely not deeply exhumed long before then. The sedimentary evidence suggests an acceleration of Greater Himalayan exhumation immediately prior to ~23 Ma.

Twenty-three million years ago is a curious age for the start of Himalayan exhumation because it greatly postdates the initial India–Asia collision at around 50 Ma. Why did the Greater Himalayas only start to form such a long time after initial compression and thrusting? Previously the 23 Ma age did not correspond to any estimates of monsoonal activity, when the emphasis was on intensification at 8 Ma. However, new evidence for strengthening monsoons in the Early Miocene in the South China Sea (Clift *et al.*, 2002a; Jia *et al.*, 2003), in the Loess Plateau (Guo *et al.*, 2002) and in the climatic zonation of China (Sun and Wang, 2005) now allows a first-order correlation between Greater Himalayan exhumation and the onset of the monsoon to be suggested. Interestingly, weathering patterns derived from geochemical proxies (e.g., the Chemical Index of Alteration; Nesbitt and Young, 1982) in the Arabian Sea appear to parallel the longer South China record during the Miocene, suggesting that the South and East Asian monsoons changed in parallel with one another and that the Himalayan front should have experienced stronger summer rains after 23 Ma.

At present, the link cannot be tested rigorously because the detailed climatic records for monsoon intensity in South Asia during that time period are lacking and until the start of Himalayan exhumation is better dated the link will remain unsure. However, what is known is that in these mid-latitudes rain is typically scarce in continental interiors and that climate models predict that the region of heavy summer rains extends gradually further inland from the coast as the monsoon strengthens in response to Tibetan surface uplift. Kitoh (2004) modeled heavy rain only reaching the Himalayan front after around 40% of the Tibetan Plateau had been formed (Figure 5.25). Most modern tectonic models for the exhumation of the Greater Himalayas appear to require erosion, driven by precipitation at the range front (e.g., Nelson *et al.*, 1996; Beaumont *et al.*, 2004). In this context it is hard to see how the Greater Himalayas could have formed without a strong monsoon, in addition to the crustal thickening caused by continental collision.

5.11   Summary

We conclude by noting that the monsoon has had a powerful influence on the erosion and weathering of Asia over long and short geological time spans during the Cenozoic. Coherent patterns of mass transport to the oceans across regions with contrasting tectonic regimes, and even into Europe, indicate a climatic rather than regional tectonic control on erosion. In Asia this climate

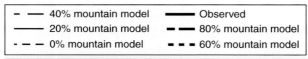

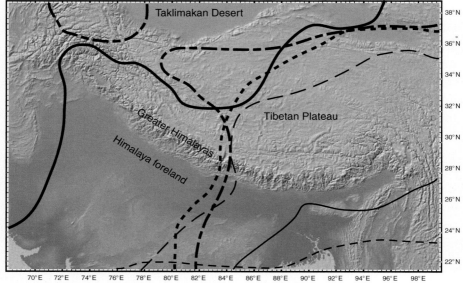

Figure 5.25 Map showing the extent of the $1.0\,\mathrm{mm\,day}^{-1}$ mean precipitation for June observed and for a series of climate models predicting the intensity of summer rains for Himalayan–Tibetan topography at different percentages of its current elevation (Kitoh, 2004). Note that the rain front only migrates significantly inland after 40% plateau elevation has been achieved.

control is manifest as monsoon intensity. A stronger summer monsoon results in faster erosion and more intense chemical weathering. The ability of summer monsoon precipitation to remove large volumes of rock from tectonically active mountains, especially the Himalayas, has resulted in an important coupling between the climate and the tectonic evolution of the mountains. A positive feedback is in operation where precipitation causes erosion and exhumation and in turn focusses further rock uplift, intensifying the situation further. It now seems that initial intensification of the monsoon in the Early Miocene first brought rains to the Himalayan front and allowed rapid exhumation of these ranges after that time. A slowing of erosion rates seen in the mountain sources and marine sediment records during the Late Miocene ($\sim$10 Ma) appears to correlate with a weakening of monsoon rains.

6

The Late Holocene monsoon
and human society

6.1   Introduction

Two-thirds of the world's population depend on the Asian monsoon to bring much needed moisture for agriculture and drinking water, the basic needs underpinning society. However, the yearly rains can also bring peril. Floodwaters triggered by summer monsoon rain often kill hundreds, sometimes thousands, of people in China, India, Nepal, Bangladesh and throughout Indochina, displacing millions of others. The high population density and corresponding competition for fertile agricultural land and water resources means that the summer monsoon rains are central to continued economic and cultural development of South and East Asia. Future changes in monsoon intensity will have enormous societal consequences, especially because the regional population is so large and the pace of economic development has been so rapid. Moreover, with four nuclear-armed countries in the modern monsoon region, understanding of the long-term stability of the monsoon has never been more important to maintaining prosperity and peace across this area (Figure 6.1).

In this chapter we review how human societies have developed across Asia since the start of the Holocene and draw some conclusions about the role that the monsoon may have played in this process. We choose to compare human cultural development with some of the more detailed climate records in order to make convincing links between the two. We employ records derived from speleothems, ice cores, lakes and peat bogs to assess how monsoon strength has varied since 8000 y BP, with an emphasis on dividing events subject to the southwest monsoon and the East Asian monsoon from each other. In Figure 6.2 we employ the speleothem cave records from Defore and Qunf Caves in southern Oman, as reconstructed by Fleitmann *et al.* (2003, 2004), to trace the intensity of

197

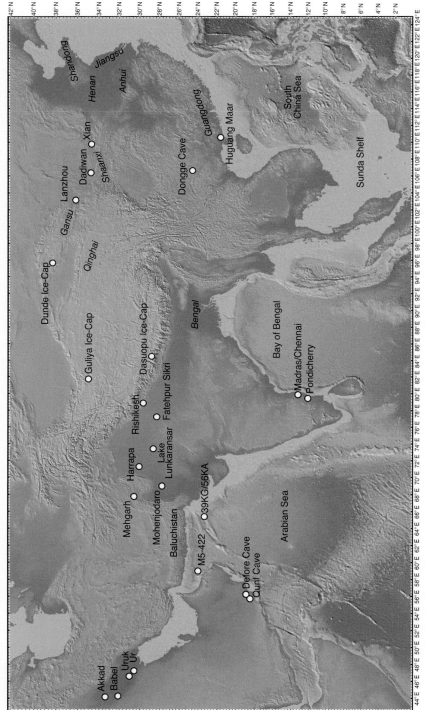

Figure 6.1 Shaded topographic and bathymetric map of Asia and the Middle East showing the location of the major sites discussed in this chapter.

198

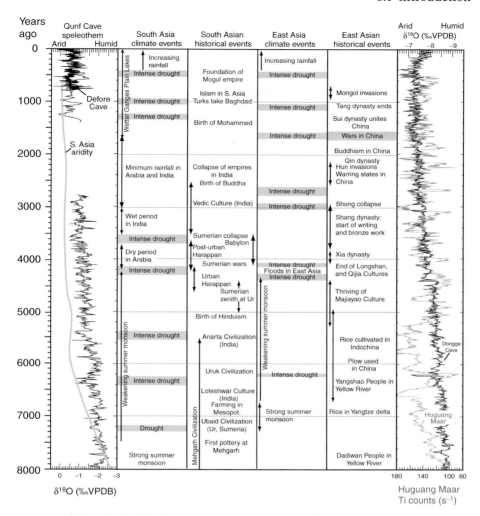

Figure 6.2 Diagram comparing historical and climatic events since 8 ka in west and east Asia. Qunf and Defore Cave isotope records from Oman are from Fleitmann *et al.* (2003) and (2004) respectively. Plot of "south Asian aridity" is from Prins and Postma (2000). Dongge Cave speleothem record is from Wang *et al.* (2005), while the Hongyuan peat bog carbon record is from Hong *et al.* (2001, 2003).

the summer monsoon in western Asia. We also compare this with the "South Asian Aridity" proxy of Prins and Postma (2000), which is based on the flux of eroded material from land into the Gulf of Oman. Together these show an overall trend to drier conditions from 8000 y BP to the present day, but with rapid changes in monsoon intensity only resolvable in the speleothem record. Further east we employ the high-resolution speleothem record from Dongge Cave in

SW China as a summer monsoon proxy (Wang *et al.*, 2005), together with the eolian titanium concentration record from lake sediment in southern China as a detailed winter monsoon tracer (Yancheva *et al.*, 2007).

The link between human cultural development and monsoon intensity is not hard to understand, as the character of any given civilization must necessarily reflect the conditions under which it was formed. At the time of the Last Glacial Maximum, ~20 ka, it is generally believed that mankind survived primarily as small population groups using basic hunter–gatherer methods to derive the bulk of their calories (e.g., Hayden, 1981; Richerson *et al.*, 2001). This life style was adapted to suit the generally dry, cold and harsh climate of the lands around the fringes of the northern hemispheric ice sheets, where vegetation was sparse and the winters were long and difficult. In this environment, sustainable, productive agriculture would have been impossible (Richerson *et al.*, 2001). However, following the Younger Dryas, around 11 ka the rapid amelioration of climate across Eurasia and North America opened up a range of new possible habitats. Likewise, in Asia the summer monsoon strengthened as the North Atlantic basin warmed, allowing vegetation to spread and diversify. The increased rains and warmer temperatures greatly increased both the area of land available for habitation and the numbers of people that any given area could sustain. The change in climate appears to have been a trigger for migrations of populations (Jin and Su, 2000; Underhill *et al.*, 2001), although it is far from clear that this was a time of great cultural complexity. There are very few ruins of settlements dating from the Early Holocene and the basic hunter–gatherer mode can be inferred to have been dominant during this period.

Ironically, it appears that the first complex human societies started to form during the Mid Holocene, after ~8000 y BP, when the climate began to desiccate and become less hospitable (deMenocal, 2001). Deteriorating climatic conditions would have forced smaller groups to work together to find sufficient food to sustain them, while the change in climate seems to coincide with the oldest evidence for organized agriculture. If food could not be gathered from wild plants then the systematic planting and harvesting of grains and fruit would have been an effective response. However, a shift to agricultural life styles necessarily requires that the community tending the fields is stationary for much of the year. Once a population had made the shift from the more nomadic hunter–gatherer life style then it would have made more sense to expend time and effort to construct more robust and thus preservable dwellings. Ruins from these early periods are largely centered in the four great river valleys of the Nile, the Tigris-Euphrates, the Indus and the Yellow (Huanghe). Clearly the monsoon would have been especially important to the Indus and the Yellow River. Some of the oldest sites have been uncovered in the Indus Valley regions of western

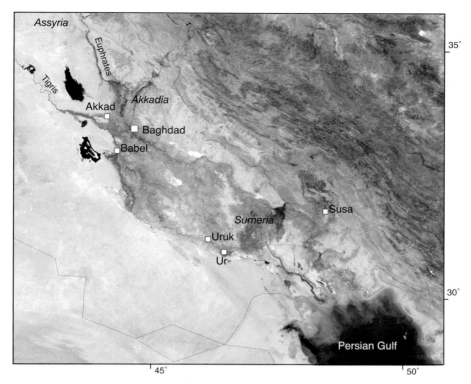

Figure 6.3 Satellite image of the Mesopotamia region from NASA's Visible Earth library showing the major archaeological sites along the Tigris-Euphrates River Valley.

Pakistan and Afghanistan and date back to around 7000 BCE (Possehl, 1990; Gupta, 2004). These agricultural communities are named the Mehrgarh Culture. The oldest cultural sites in eastern Asia are from the Dadiwan Culture of the middle Yellow River drainage in what is now western Gansu and Shaanxi provinces of China, which date from ~7800 y BP (An *et al.*, 2004, 2005). This coincides with the near synchronous emergence of agriculture along a vast east–west transect centered on the Yellow and Wei Rivers in northern China (Bettinger *et al.*, 2007; Ren, 2000). At around the same time agricultural communities were recognized in the lower Tigris-Euphrates Valley, with the development of the Ubaid Civilization in Sumeria, close to the city of Ur (Young, 1995; Figures 6.1 and 6.3).

As climate in Asia continued to dry, the pressure to form larger cities in preference to smaller villages would have increased. A more arid climate would not necessarily immediately result in societal collapse, but would drive communities to integrate and organize to help themselves and their neighbors. In doing so the drying climate would have precipitated the development of the first city-states, such as at Ur in Sumeria and at Harappa in the Indus Valley (Figure 6.4).

Figure 6.4 (a) Photograph by Lasse Jensen of the Great Ziggurat at Ur, Sumeria, modern Iraq. (b) Image of an Indus Valley Civilization excavation site of the ruins of Mohenjo-daro, Pakistan. The prominent feature presented in this image is called the "Great Bath" (published with permission of Saint Xavier University, Chicago, Illinois). See Figure 6.1 for locations.

Where sustainable agriculture required the construction and maintenance of a complex irrigation system communities would be drawn firstly to the source of water, generally the major, monsoon-fed rivers of Asia. Greater cultural organization would be required because if a proportion of the population was employed in maintaining the irrigation channel networks then these same people could not be tending fields or preparing food. Specialization of labor can only develop in such complex societies consisting of many individuals in a single organized community (Tosi, 1975; Peregrine, 1991). Similarly, the arts and sciences can only be pursued by a sub-section of society when others are doing the work of everyday subsistence. This is only practical in the context of the city-state that was brought into being by the weakening of the summer monsoon in Asia after 8000 y BP. We now look at a number of specific examples of how rapid change in the monsoon climate in the late Holocene drove cultural development in Asia.

6.2   Holocene climate change and the Fertile Crescent

The flood plains of the Tigris-Euphrates Rivers, regions often referred to as the "Fertile Crescent", lie just west of the Asian monsoon system and are widely recognized as home to some of the oldest organized urban cultures on Earth. Starting with the Ubaid Culture ~7000 y BP, the remains of a series of civilizations have been identified from the region. The city of Ur was inhabited during this Ubaid period, when it formed one of a series of village settlements in southern Mesopotamia (Figure 6.3). However, as the climate changed from relatively moist to drought-prone after 5000 y BP (Figure 6.2), paralleling the monsoon, the small farming villages of the Ubaid Culture consolidated into larger settlements. This trend arose from the need for large-scale, centralized irrigation works that were required for cultivation under increasingly harsh climatic conditions. Ur became one such urban centre, and by around 4600 y BP the city was thriving. The first phase of urban development at Ur (first dynasty) was ended by an attack from the Akkadian empire to the North around 2340 BCE (4340 y BP; Reade, 2001). This was a time of particularly dry climate, which may have weakened the city prior to the final Akkadian military assault. Ur remained a minor regional player for around 200 years after this event, but regenerated after 2112 BCE.

The Ziggurat is the most prominent remaining building at Ur and was the temple of Nanna, the moon deity in Sumerian mythology (Figure 6.4(a); Albright, 1944). The temple was constructed along with several other structures around 2100 BCE. The city never fully recovered from another attack in 1950 BCE, this time from the Elam people, based around the city of Susa, in what is now SW Iran. The Elam Culture itself is an ancient civilization, dating from around 3200 BCE. It is believed that Susa began to thrive after a migration from the Iranian plateau to the east, where increasing aridity made life outside the largest city-states harder (Potts, 1999).

Around 4200 y BP the ancient civilizations of Mesopotamia, most notably the Akkadian Empire (Figure 6.3) rapidly contracted or even collapsed entirely (Weiss *et al.*, 1993; Hassan, 1997). Records show that this decline was triggered in part by drought. To assess whether this drought was linked to a regional climatic shift, Cullen *et al.* (2000) used a deep-sea core from the Gulf of Oman to examine

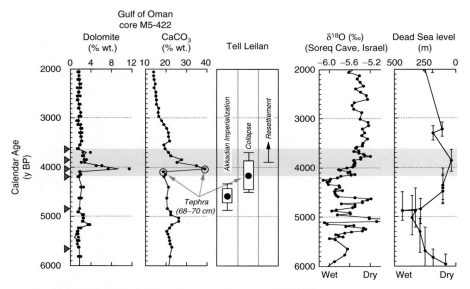

Figure 6.5 Eolian mineral concentration data for 6000–2000 calendar y BP interval in core M5-422 from Cullen *et al.* (2000). Calibrated radiocarbon ages of imperialization, collapse and resettlement phases of Akkadian empire as determined from archeological investigations at Tell Leilan, Syria, are provided by dates from Weiss *et al.* (1993). Mean calibrated ages of these phases, their 2σ and full age ranges, are represented by filled symbols, boxes and range bars, respectively.

changes in the oceanographic and atmospheric conditions. Around 4200 y BP there was a dramatic increase in the amount of eolian dust accumulating in the ocean, which geochemical and mineralogical data indicate to have been carried from the west (i.e., from Sumeria and Akkadia) into the deep-sea basin (Figure 6.5). The ^{14}C ages derived from the carbonate shell material further allowed these workers to demonstrate that the ultra-dry period lasted for ~300 y. This period of drying out and strong dust storms coincided with historically low water levels in the Dead Sea, following a high at ~5000 y BP (Frumkin, 1991), giving confidence that the societal collapse at this time was in response to a regional climatic event.

Because of the temporal coincidence between climate change and steep decline of the urban centers it has been inferred that around 4200 y BP climatic desiccation had reached the point where even the organized structures of the city-state, originally developed to combat the increasing aridity of the mid-Holocene, were unable to continue to sustain a large population in a single community. It is not hard to imagine that as years of drought followed each other the outlying villages of the Akkadian Empire were unable to sustain themselves, driving the people into the larger cities, and thus stressing the ability of the remaining parts of the centralized state still further. If droughts had not been long lasting then it is probable that the empire would have been able to recover, yet the 300 years of ultra-dry climate experienced proved too much and appears to have been the underlying cause of collapse of the Akkadian cities, as well as those at Ur and in the Nile Valley, followed by a dispersal of the population.

Nonetheless, it is noteworthy that after this, 4200–3900 y BP dry spell and a return to more normal climatic conditions new city states were able to establish themselves in the region. This was especially true in the resurgent state of Babylon (Figure 6.3). After this time the Babylonians extended their rule and engaged in regular trade and influence with Western city-states. Babylonian officials and troops were known to have passed into Syria and the states of the eastern Mediterranean. By 3780–3750 y BP Babylon had become the most important city in Mesopotamia, and indeed the whole Middle East. It has been estimated that Babylon was the largest city in the world from 3770 to 3670 y BP and was perhaps the first city to reach a population above 200 000. The Babylonian culture itself collapsed in 3595 y BP, again at a time of weak summer monsoon intensity, as recorded in the Oman speleothems (Fleitmann et al., 2003; Figure 6.2). Thus we have evidence that climatic changes, linked to the south-west Monsoon, controlled the timing and nature of cultural development in the Fertile Crescent during the earliest phases of permanent human settlement.

6.3   Holocene climate change and the Indus Valley

Remains of the earliest human settlements in the Indian subcontinent are all found in what is now western Pakistan and eastern Afghanistan, and are clearly related to the establishment of small-scale farming communities. These settlements are known as the Mehrgarh Culture after the best-known site, close to modern Quetta, Pakistan (Figure 6.1). Mehrgarh is sometimes cited as the earliest known farming settlement in South Asia, based on archaeological excavations that were conducted from 1974 to 1986 by a French team (Jarrige, 1993). Related remains are also found in Baluchistan in SW Pakistan, on the plains near the Bolan Pass, to the west of the Indus River valley and between the present-day cities of Quetta, Kawas and Sibi (Figure 6.6). The earliest Mehrgarh remains date from 7000 BCE (9000 y BP) and were simple Neolithic in character, showing no production of pottery (Lal, 1997). Nonetheless, these settlements appear to have been in contact with western Asia and together formed part of the "Neolithic Revolution" seen in the Fertile Crescent around 9000 to 6000 BCE. The earliest farming in the area was developed by semi-nomadic people using wheat and

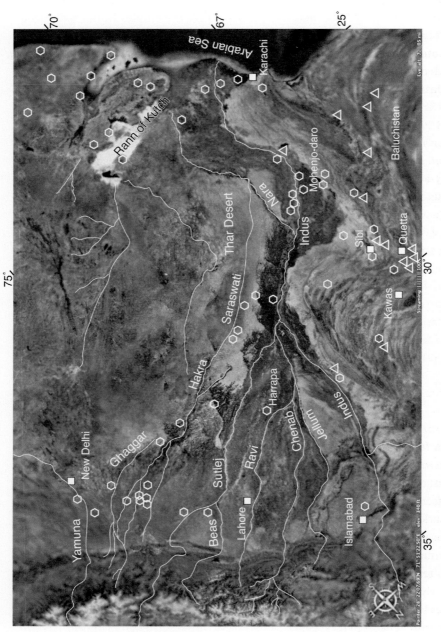

Figure 6.6 Satellite map of the Indus drainage system showing the modern river and associated tributaries, together with the course of the former Saraswati, now reduced to the seasonal Ghaggar River in India and the abandoned channel of the Hakra in Pakistan. Hexagons show known locations of major Harappan sites. Triangles show locations of early farming Mehgarh culture sites. The map is modified from *TerraMetrics*, Google Earth™.

barley, supplemented by the herding of sheep, goats and cattle. The settlements themselves were basic, comprising simple mud buildings with four internal subdivisions, collected into small communities.

Pottery was one of the first technical innovations of these peoples and was first noted to be in use by around 5500 BCE (7500 y BP). With time, more sophisticated casting and firing methods were developed, including the painting and glazing of ceramics that were placed in burial pits (Possehl, 1993). Toward 5500 y BP, the amount of material buried along with their owners decreased, suggestive of a society in decline and under economic pressure, possibly due to environmental stress as the summer monsoon continued to weaken. Between 2600 and 2000 BCE Mehrgarh and the related settlements seem to have been largely abandoned, coinciding with the early development of the Indus Valley Civilization in the floods plains of the Indus River to the east. It seems likely that the inhabitants of Mehrgarh and associated communities migrated to the fertile Indus valley as Baluchistan became more arid during weakening of the summer monsoon rains.

6.3.1   The Harappan Civilization

The Indus Valley (Harappan) Civilization grew out of the Mehrgarh culture's technological base. The Harappan civilization was clearly very advanced, with cities laid out in orderly grid patterns and with a well organized water and waste disposal system. Harappa and Mohendojaro's plans included the world's first urban sanitation systems (Figure 6.4(b)). Within the cities, individual homes or groups of homes obtained water from wells. Within individual houses a room was set aside for bathing, with waste water directed to covered drains, which lined the major streets. The Harappan culture also developed written records. The earliest examples of the Indus script date from this period, placing the origins of writing in South Asia at approximately the same time as those of Ancient Egypt and Mesopotamia (e.g., Shaffer, 1992). By 2600 BCE (4600 y BP), it had risen to become a complex civilization. The agriculture of the Harappan civilization must have been highly productive, as it was capable of generating surpluses sufficient to support tens of thousands of urban residents who were not primarily engaged in agriculture. The Harappan system relied on the considerable technological achievements of the pre-Harappan culture, including the plow. The efficiency of the agriculture reflected the growing environmental stresses linked to a weakening monsoon that required groups of people to work together to generate sustainable food production.

However, the urban and supporting agricultural systems of the Harappan culture changed after 4200 y BP, when the civilization transformed from this early highly organized urban phase to a post-urban phase of smaller settlements

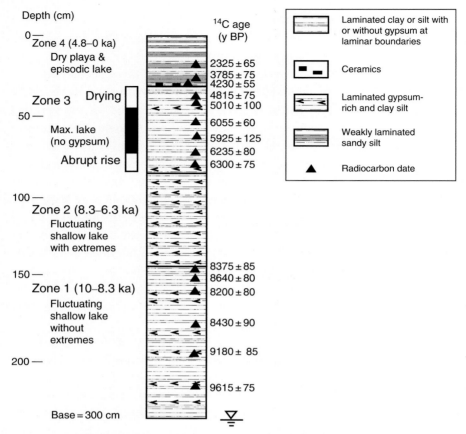

Figure 6.7 Four stratigraphic zones of the Holocene deposits of Lake Lunkaransar (from Enzel *et al.*, 1999). Note the rapid facies change to a more evaporating environment between 4815 and 4230 y BP. The lake location is shown in Figure 6.1.

accompanied by a general south-eastward migration of the population (Possehl, 1997). Exactly what precipitated this change is not totally clear, but it is note-worthy that this collapse coincided with the one seen in Mesopotamia, sugges-tive of an environmental trigger. Climatic reconstructions made from marine and terrestrial records from the region now confirm this hypothesis. Cores taken from Lake Lunkaransar in north-western India in the Thar Desert (Figure 6.1) show that prior to 4815 y BP the lake accumulated laminated silts and clays (Figure 6.7). Although there were rises and falls of lake level, the lake was clearly a permanent feature of the area. However, these same cored sediments now show that after 4230 y BP, the lake was only temporarily filled with water and was often reduced to being a saline playa (Figure 6.7; Enzel *et al.*, 1999). Such

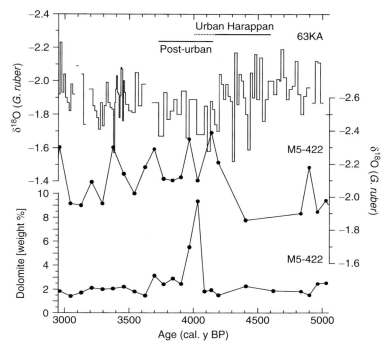

Figure 6.8 Plot from Staubwasser *et al.* (2003) showing the link between monsoon-driven climate change and human civilization in Southwest Asia. Core 63KA (north-eastern Arabian Sea) $\delta^{18}O$ of *Globigerinoides ruber* compared with core M5-422 (Gulf of Oman) d^{18}O (*G. ruber*) and dolomite concentration (Sirocko, 1995; Cullen *et al.*, 2000). Dolomite is a proxy for dust flux from a northern Arabian source. The same ^{14}C reservoir correction was used in both cores. Harappan and Akkadian cultural phases are given for comparison.

a change is interpreted to indicate rapid weakening of the SW monsoon rains, consistent with what is known from the speleothem record of southern Oman (Fleitmann *et al.*, 2003) and from pollen studies (Herzschuh, 2006).

Additional corroborating evidence for regional desiccation comes from the offshore of Pakistan. Staubwasser *et al.* (2003) used cored sediment recovered from close to the Indus delta mouth (Core 39KG/56KA; Figure 6.1) to examine changes in the oxygen isotope character ($\delta^{18}O$) of planktonic foraminifers that lived in the upper parts of the water column at this time. Surprisingly this core yielded a record with the opposite trends to those seen at core M5-422 in the Gulf of Oman (Figure 6.8; Cullen *et al.*, 2000). Although $\delta^{18}O$ can be controlled by changes in sea-surface temperatures this seems unlikely because the two core sites are so close to one another. In addition, a Late–Mid-Holocene alkenone record from the north-eastern Arabian Sea does not indicate significant sea-surface temperature changes between 4500 and 3500 y BP (Doose-Rolinski *et al.*, 2001). Staubwasser *et al.* (2003) inferred that the change in $\delta^{18}O$ they identified was driven by a salinity change in the ocean near to the Indus delta and that water discharge from the Indus River slowed rapidly at 4200 y BP, consistent with other regional data for climate change at that time (deMenocal *et al.*, 2000). Such a change is interpreted to indicate rapid weakening of the SW monsoon rains, consistent with what is known from the speleothem record of southern Oman (Fleitmann *et al.*, 2003; Burns *et al.*, 2002).

6.3.2   The Saraswati River

Competing theories have been advanced to explain the demise of the Harappan urban culture after 4200 y BP. Drainage patterns in river catchments are known to vary in space and time in response to changes in climate, sediment accumulation rates and tectonic forces. Such changes in drainage patterns can have major implications for human civilizations that inhabit the river plain. Dramatic changes in major rivers are known from recent times, such as the repeated switching of the Yellow River mouth from south to north of the Shandong Peninsular since 1194. The last switch occurred as recently as 1897. The Indus River has experienced a number of major drainage re-organizations since its inception during the Eocene, most notably the capture of the Punjabi tributaries from the Ganges after 5 Ma (Clift and Blusztajn, 2005). More recently the drainage of the western Himalayas appears to have lost a major independent river, known as the Saraswati, which used to flow out of the Himalayan mountain front between the modern courses of the Yamuna and the Sutlej (respectively the western and easternmost tributaries to the Ganges and Indus). The Saraswati then ran sub-parallel to the Indus toward the Arabian Sea according to descriptions from the Sanskrit Hindu holy text, the Rig Veda, in which the river is mentioned 72 times (Figure 6.6).

The Rig Veda was written ~3500 BCE (~5500 y BP) and described the Saraswati as being a major river, perhaps comparable in size to the Indus, sourced in the Greater Himalayas and flowing to the Arabian Sea near the Rann of Kutch (Oldham, 1893; Murthy, 1980). Such a river does not exist today, but satellite photography has been used to chart a number of abandoned channels or seasonal rivers that start in the Himalaya and terminate in the Thar Desert (e.g., Drishadvati and Ghaggar Rivers; Ghose *et al.*, 1979; Kar, 1984; Puri, 2001). The potential importance of this river to early civilizations of the region is hard to overstate. Nearly 2000 of the 2600 Harappan sites that have been discovered are proposed to have lain on the old paleo-channel of the Saraswati and its presumed outwash in the Rann of Kutch (e.g., Gupta, 1995). If capture of the Saraswati by the modern Indus drainage did occur then this is not to say that changes in monsoon intensity were not involved in the process of societal collapse. The role of drainage capture in controlling the fate of early civilizations in western India and Pakistan cannot be properly tested until the existence of a former Saraswati River is demonstrated and dated through sampling of former channel sediments.

6.4   Holocene climate change and early Chinese cultures

As in western Asia, paleoclimate records in China point toward more humid conditions in the Early Holocene, followed by a long-term drying after ~8000 y BP (Figure 6.2). Carbon isotope data from the Loess Plateau indicate C4 abundance increased from the last glacial to the Holocene under the influence of a stronger summer monsoon (Liu *et al.*, 2005). Similarly, climatic variability immediately before, during and immediately after the Younger Dryas cold climatic interval (~11 ka) is associated with rapid technological elaboration and innovation in the production and use of chipped stone tools, and perhaps, ground stone (Madsen *et al.*, 1998). This climatic variability, which is associated with the end of the Younger Dryas, appears to be linked with a transition to broad-spectrum foraging and seed processing by hunter–gatherers in western and central China, and, ultimately to the advent of agriculture. Agriculture and the establishment of permanent settlements in China is closely temporally associated with the drying of the climate, i.e., weakening of the summer monsoon. The earliest human settlements in China are associated with the Yellow River valley in modern Gansu and Shaanxi provinces of northern central China, between the cities of Xian and Lanzhou (Figure 6.1).

Coring and field sampling of Holocene sediments from this part of the Loess Plateau allows the evolving environment to be reconstructed and correlated with archaeological evidence. The region today is marked by a semi-arid to increasingly arid climate, moving west into the Tengger and Mu'us Deserts (Figure 3.22). However, these conditions are apparently relatively new to the region. Feng *et al.* (2004) used cores from this area to show the development of a widespread wetland and swamp layer throughout the region that dated from 10 000 y BP to as young as 4000 y BP. In particular, this work identified a series of swamp and river sediments that contained abundant aquatic mollusk shells, dating from 8000–6000 y BP at Dadiwan in north-east China (Figures 6.1 and 6.9) indicating wet conditions, followed by drying at 6000–5000 y BP (Huang *et al.*, 2000). This is significant because Dadiwan is the location where the earliest human settlement remains have been identified. Recent work using the alkenone temperature proxy on sediments cored from Lake Qinghai confirm that there has been a close relationship between average surface temperatures and lake salinity changes during the Late Holocene (Liu *et al.*, 2006). Lake salinity is largely

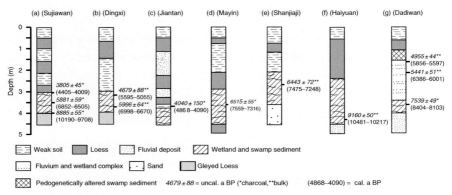

Figure 6.9 Strata and C-14 dates of the wetland and swamp layer at the eight focal sites in the western part of Chinese Loess Plateau, from Feng *et al.* (2004). The presence of this layer indicates humid conditions in northern China prior to ∼4000 y BP followed by progressive desiccation.

controlled by summer monsoon precipitation and the climate record showed a stronger monsoon during the Medieval Warm Period, but weaker intensities during the Little Ice Age.

The oldest known civilization from the Yellow River valley is known as the Dadiwan culture, a Neolithic civilization that existed along the central Yellow River in China at 5800–5350 BCE (Figure 6.10). This culture was in turn succeeded in the same area by the more extensive Yangshao culture, which is dated from around 4800 BCE to 2900 BCE (6800–4900 y BP; Elston *et al.*, 1997). The culture is named after Yangshao, the first excavated village of this culture, which was discovered in 1921 in Henan Province, although the archaeological site of Banpo village, near Xian, is the best-known site related to this cultural type. At the same time (4100–2600 BCE) the Dawenkou culture was known from further east, largely in Shandong province and the coast south of that peninsula. The Yangshao culture flourished mainly in the provinces of Henan, Shaanxi and Shanxi, although settlements were far more numerous after 4000 BCE, mirroring the rapid weakening of the monsoon and drying of the climate at this time that drove populations to work as larger groups to maximize agricultural production (Figure 6.10). There is abundant evidence that Yangshao people were agriculturalists, cultivating millet, wheat and rice, and employing the plow no later than around 4000 BCE (6000 y BP). The Yangshao kept an array of domestic animals, but much of their meat is believed to have come from hunting and fishing. Compared with the earlier Dadiwan culture, the Yangshao showed greater technical skills and cultural sophistication. Their stone tools were polished and highly specialized. The Yangshao people may also have practised an early form

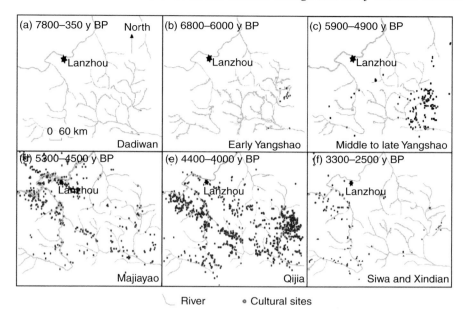

Figure 6.10 Temporal distribution of the unearthed archaeological sites in the western part of the Chinese Loess Plateau from An *et al.* (2005). (a) Dadiwan Culture (7800–7350 cal. y BP); (b) early Yangshao Culture (6800–6000 cal. y BP); (c) middle to late Yangshao Culture (5900–4900 cal. y BP); (d) Majiayao Culture (5300–4500 cal. y BP); (e) Qijia Culture (4400–4000 cal. y BP); (f) Siwa Culture (3300–2500 cal. y BP) and Xindian Culture (3600–2600 cal. y BP).

of silkworm cultivation. The Yangshao and Dawenkou cultures were well known for their painted pottery (Figure 6.11(b)). Yangshao craftsmen decorated their works with detailed white, red and black paintings that depicted human facial, animal and geometric designs.

The Yangshao culture was subsequently succeeded by the Majiayao culture after ∼3300 BCE (5300 y BP). The Majiayao are represented by a prolific series of Neolithic remains, primarily located in the upper Yellow River region in Gansu and Qinghai provinces, to the north-west of the center of Yangshao settlement. These sites are best known for hosting the earliest discoveries of copper and bronze objects in China (Shui, 2001). The Majiayao culture existed until 2400 BCE, and were succeeded by the even more numerous and technically advanced Qijia culture (2400–2000 BCE; An, 1991). These cultural transitions were all made against the background of a weakening summer monsoon. The Qijia people formed an early Bronze Age culture that was distributed around the upper Yellow River region of western Gansu and eastern Qinghai (Higham, 1996). The Qijia culture produced some of the earliest bronze and copper mirrors found in

Figure 6.11 (a) Gui from Dawenkou culture. Photograph taken at Peking University, China. Dawenkou culture was primarily developed in Shandong, but also appeared in Anhui, Henan and Jiangsu, China. The culture existed from 4100 BCE to 2600 BCE, co-existing with the Yangshao culture. (b) A painted pottery pan basin, Taosi Longshan Culture (2500–1900 BCE) © The Institute of Archaeology, Chinese Academy of Social Sciences, Beijing.

China, and evidence for the extensive domestication of horses is found at many Qijia sites. At the same time as the Qijia were thriving in the upper Yellow River valley, a second late Neolithic culture was developing in the central and lower Yellow River in China. This population is known as the Longshan culture and their remains have been found in and around the Shandong Peninsula (Figure 6.1).

It is noteworthy that evidence from bone and dental remains indicates that onset of the cooler, drier climate after 3900 y BP marks the decline of the egalitarian society of the Yangshao and the rise of the more divided classes in a chiefdom-like society of the Longshan (Pechenkina *et al.*, 2001). A distinctive feature of Longshan culture was the high level of skill in pottery making, including the use of pottery wheels. Longshan culture was noted for its highly polished black or eggshell pottery (Figure 6.11(a)). Life during the Longshan culture marked a transition to the establishment of cities, as rammed earth walls and moats began to appear. Rice cultivation was clearly established by that time, interpreted as a response to the increasing environmental stresses of a weaker summer monsoon (Huang *et al.*, 2000). Despite this the Neolithic population in China reached its peak during the Longshan culture.

The Longshan culture was founded about 3000 BCE but like the Qijia died out around 2000 BCE (Liu, 1996). Toward the end of the Longshan period, the population decreased sharply; this was matched by the disappearance of high-quality, black pottery found in ritual burial sites (Underhill, 1991). Similarly, during the late stages of the Qijia culture, settlements retreated from the west and suffered a reduction in population size. The density of settlements known in the Yellow River valley after ~2000 BCE, such as during the Siwa and Xindian civilizations, is much lower and indicative of less hospitable conditions than those seen during the Qijia and Longshan times (Chen and Hiebert, 1995). Clearly the land was not able to support the numbers that had existed prior to 2000 BCE.

These population trends are consistent with the hypothesis that monsoon intensity was a primary control on human cultural development, as this time was one of significant summer monsoon weakening (Figure 6.2). However, it is noteworthy that speleothem and bog climate records both show a rapid shift to drier conditions after ~4200 y BP (2200 BCE), which appears to be slightly earlier than the apparent extinction ages of the Qijia and Longshan cultures. Whether this time delay is real or whether this reflects poor dating of the archaeological sites is unknown. The fact that both the Qijia and Longshan showed sharp population reductions toward the end of their span suggests that the dating is correct and that the settlements were able to survive for as long as 200–300 years after the climate change, albeit in a reduced form. It is also significant that the oldest known imperial dynasty to control larger areas of eastern China, the Xia dynasty, dates from ~2000 BCE, immediately following the collapse of the Qijia and Longshan cultures under the stresses of droughts and then floods around 1900 BCE (Pang, 1987; Figure 6.2). It is possible that the Xia were able to prosper by moving into areas of opportunity vacated by the earlier groups, and to cope well with the deteriorating climate with greater centralized governmental organization.

Further detailed correlation of climate change and societal evolution in East Asia has been possible using a record of winter monsoon activity derived from sediments cored from Huguang Maar in Guangdong Province (Figure 6.1). Yancheva et al. (2007) used Ti concentrations as a proxy for winter monsoon wind strength, because this element is enriched in the eolian dust brought to the lake during the winter. Figure 6.2 shows the Ti record from Huguang Maar that can be compared with historical events dating back 8000 years. The collapse of several Chinese dynasties appears to correlate with periods of strong, dry winter monsoons, e.g., the Tang Dynasty at 907 AD, the Qin Dynasty at 206 BCE and the Shang Dynasty at 1046 BCE. It is also noteworthy that the period of Warring States correlates with an extended period of dry, windy conditions at 475 BCE to 221 BCE.

The climatic evidence from Huguang Maar is supported by a loess dust record from the Loess Plateau (Huang et al., 2003), which shows increased climatic aridity at 1150 BCE (3150 y BP). These conditions triggered a severe environmental deterioration in that region, resulting in degradation of natural resources. Water shortages caused by decreases in precipitation and deficits in soil moisture resulted in poor harvests and great famines that caused domestic upheavals, population migrations and even conflicts between populations that contributed to the end of the Shang dynasty (Loewe and Shaughness, 1999; Huang *et al.*, 2003). As in western Asia the shift to a weaker monsoon after 8000 y BP first stimulated cooperative agricultural and urban communities to form, then progressively caused them to collapse as the climate made food production for large groups unsustainable, especially after ~4200 y BP.

6.5   Monsoon developments since 1000 AD

The interaction of human society and monsoon climate can be examined in greater detail over the last 1000 years because of the superior written historical record and in the last 200 years by a more accurate record of changing climatic conditions through the region. The most detailed climatic record available for the period comes from ice-core analysis, which covers western and north-eastern Tibet and the Himalayas (Thompson *et al.*, 1993, 1997, 2000). These reconstructions can be supplemented by the speleothem record from Defore cave in Oman (Figure 6.12; Fleitmann *et al.*, 2004) and by tree-ring records from across the region (e.g., Liu *et al.*, 2004b; Hughes *et al.*, 1994; Sheppard *et al.*, 2004; Treydte *et al.*, 2006) (Figure 6.13). The longest oxygen isotope profile derived from Defore Cave clearly shows the transition at ~1320 AD from a generally wetter Medieval Warm Period to a drier Little Ice Age that lasted from approximately 1320–1660 AD in Southern Oman. The decrease in monsoon rainfall in the late twentieth century is also obvious in meteorological records from Arabia and north-western India, indicating that the speleothem-based rainfall records do not only reflect local monsoon rainfall variability.

6.5.1   Ice-core records

Over the timescales considered here, the $\delta^{18}O$ record from ice cores can be interpreted as largely reflecting temperature variations across the Tibetan Plateau that, in turn, are linked to the varying strength of the summer monsoon. All three ice cores from Tibet (Figure 6.12) demonstrate a large-scale twentieth century warming trend that appears to be especially amplified at higher elevations (Thompson *et al.*, 2000). However, the shift to a warmer climate with more positive $\delta^{18}O$ can be seen to date back to the start of the nineteenth century. This isotopic warming is the most regionally consistent climate signal of the last 1000 years. However, over most of the past millennium, the temperature

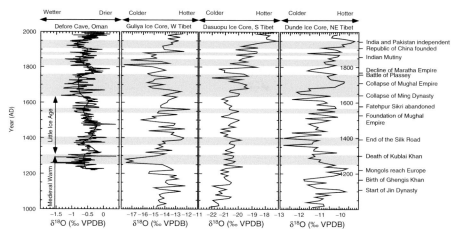

Figure 6.12 Diagram showing four different proxies for monsoon intensity over the last 1000 years from west to east across Asia. Data from Defore Cave, Oman are from Fleitmann *et al.* (2004). Data from Guliya ice core are from Thompson *et al.* (1997). Data from Dasuopo ice core are from Thompson *et al.* (2000). Data from Dunde ice core are from Thompson *et al.* (1993). Gray-shaded bands show periods of enhanced humidity recorded in the Defore Cave record. Note the disimilarity of the record at other study locations, indicating different histories to the monsoon in different parts of Asia. The locations of points are shown in Figure 6.1.

variations inferred from $\delta^{18}O$ values recorded from the Guliya ice cap in western Tibet (Thompson *et al.*, 1997) have been generally out of phase with those from Dunde in north-eastern Tibet (Thompson *et al.*, 1989, 1993), suggesting different large-scale climate regimes on the north-eastern and north-western sides of the Tibetan Plateau.

Dasuopu, located in the Himalayas in southern Tibet (Figure 6.1) shows the clearest overall long-term rise in temperature and differs sharply from the other two Tibetan records (Figure 6.12). The deeper, older sections of the Dasuopu core suggest that several periods of drought have afflicted this region in the last 1000 years. If $\delta^{18}O$ in ice at Dasuopu is a proxy for temperature in the Himalayas, then its strong correlation with wind-blown dust concentrations also preserved in the ice implies a direct link between warming and increased dust in the South Asian atmosphere. Dust would be generated more efficiently if there were a reduction in snow cover, increased aridity, or increased agricultural activity in the source regions. Enhanced anion concentrations may also be attributed to increased land use and widespread biomass burning for cooking and heating, as well as the increased industrialization in Nepal and India during the twentieth century (Thompson *et al.*, 2000). The most prominent drought phase seen in South Asia is dated at 1790 to 1796, which is marked by high dust and Cl concentrations, as well as by enriched $\delta^{18}O$ values. The twentieth-century increase in anthropogenic activity in India and Nepal, upwind from this site, is recorded by a doubling of chloride concentrations and a fourfold increase in dust.

6.5.2   Tree-ring records

Some of these temporal climatic patterns are mirrored in the record of tree rings, in which broad annual rings are developed in years with a moist, fertile climate, while drought years correspond to thin growth rings. Treydte *et al.* (2006) used tree-ring data from the Karakoram region of NW Pakistan to examine rainfall variations there since 1000 AD. This record was then compared with similar records from East Asia (Sheppard *et al.*, 2004) and Europe (Wilson *et al.*, 2005), as well as with monsoon-related, oceanic upwelling records from the Arabian Sea (Anderson *et al.*, 2002). As with the ice cores it is possible to see that the East and South Asia climates do not vary in phase with one another over these timescales (Figure 6.13(a)–(c)). There is even a clear difference between the Arabian Sea and the Karakoram, reflecting the influence of westerly jets in controlling the climate of the latter region. As with the ice cores there is a sharp, coherent change in the latter part of the nineteenth and twentieth centuries. However, unlike the shift to a weaker summer monsoon seen at Defore Cave, the tree records show a consistent change to heavier summer precipitation, as might be expected in the general setting of a warming global climate. This difference likely indicates a northward migration of the summer location of the Intertropical Convergence Zone. Similarly, and like the Defore speleothem record, precipitation is seen to fall in the Karakoram during the Little Ice Age, but to be variable and generally higher during the Medieval Warm Period. There is a general positive correlation between these climatic trends and those in East Asia, as well as with several measures of global average temperature (Figure 6.13(f)), albeit with modest temporal offsets between the different regions.

6.5.3   Historical correlations

Comparison of the monsoon and climate records since 1000 AD with major historical events in Asia yields some interesting correlations and potential relationships. Generally heavier monsoon rains in Asia accompanied the spread of the Mongol Empire at the end of the Medieval Warm Period. The greatest extent of Mongol conquests in Eastern Europe is dated at 1227, just prior to the end of the wet period. Indeed, as the climate dried in the early fourteenth century, the Mongol empire began first to fragment and then contract. The Silk Road, the trading artery that linked the empire from east to west was effectively closed by 1400. As the Mongols both fought and traded by horseback through

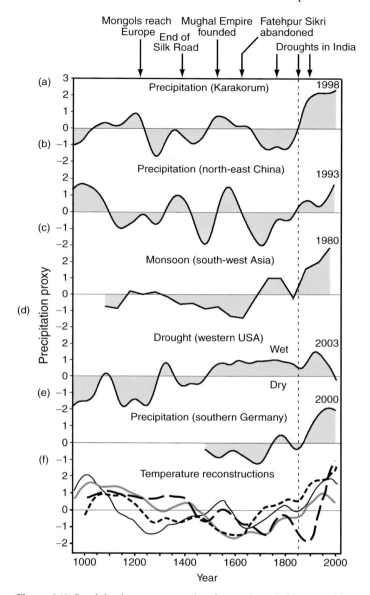

Figure 6.13 Precipitation reconstruction for northern Pakistan and long-term precipitation variations for different regions in the northern hemisphere from Treydte *et al.* (2006). (a) Tree-ring δ^{18}O-derived reconstruction. (b) Annual precipitation reconstruction (July–June) from tree rings in north-east China (Sheppard *et al.*, 2004). (c) South-west Asian monsoon intensity from *Globigerina bulloides* in the Arabian Sea (Anderson *et al.*, 2002). (d) Drought reconstruction from tree rings in western USA (Cook *et al.*, 2004). (e) Spring–summer precipitation reconstruction from tree-rings in southern Germany (Wilson *et al.*, 2005). (f) Regional to hemispheric temperature variations according to Esper *et al.* (2007) (short dashed line, western Central Asia),

central Asia it is logical that the weakening of summer rains and the contraction of fertile grassland that formed the pasture for their mounts would adversely affect the ability of the organization to function as a coherent body over such wide expanses of territory. This is not to say that climate change alone undermined the workings of the empire, as there were numerous external threats and internal political conflicts that also played a part in terminating the greatest contiguous empire the world has ever seen (Prawdin, 2006). It seems most likely that monsoon weakening would have been one of several factors that helped destabilize the central government and lead to its collapse.

The rise and fall of the Mughal Empire in South Asia also shows interesting correlations with monsoon strength. The first Mughal emperor, Babur, was installed in 1527 (Richards, 1996; Alam and Subrahmanyam, 1998), at a time when the summer precipitation was increasing, and when a strong summer monsoon would have made the Indian plains fertile and productive. The empire is generally considered to have reached its zenith under Akbar (1556–1605), although already by this time the monsoon was again weakening. In 1585 the lavish and ornate city of Fatehpur Sikri, located near Agra in northern central India, was abandoned after being completed only 15 years earlier in 1570 (Figure 6.14(a)). The city was never used again because of problems with water supply away from the Yamuna River that supplied Agra itself. A weakening monsoon would have exacerbated the water supply problems that were inherited from the location of the site. Although Shah Jahan (1627–1658) is one of the best-known Mughal emperors thanks to his architectural legacies in creating Delhi, the Red Fort and the Taj Mahal, the power of the empire was already weakening under his rule. The Mughal Empire did last in name until the middle of the nineteenth century, yet to all intents and purposes the last effective ruler was Aurangzeb (1658–1707), whose reckless policies of taxation and military adventure resulted in rapid deterioration in the empire's fortunes, no doubt aided by the stresses generated by weaker summer rains (Richards, 1990).

It was in the declining years of the Mughal Empire that European powers first started to gain colonial footholds in India, with the British establishing a post in Madras (now Chennai) in 1639 under Shah Jahan and the French at Pondicherry

Caption for Figure 6.13 (*cont.*)

Mann *et al.* (1999) (long-dashed line), Esper *et al.* (2002) (black) and Moberg *et al.* (2005) (gray). Records are normalized over their individual periods and smoothed using 150 year splines. Numbers in (a)–(e) refer to the last year of the records, and the black dashed line to the shift from negative to positive precipitation anomalies in the Karakorum record.

Figure 6.14 (a) Photograph of the Diwan-i-Khas, part of the palace complex at Fatehpur Sikri, built by the Mogul Emperor Akbar near Agra in central northern India and completed around 1571. The whole city was abandoned shortly after completion in 1585 owing to drought stresses. (b) Photograph of Hindu temple complex on the Ganga River, just north of the holy city of Rishikesh.

in 1672. It is interesting to correlate the increasing influence that European powers exercised over the local Indian rulers as the monsoon continued to weaken. As the agricultural base of Mughal and then Maratha imperial power was progressively weakened, local rulers found themselves increasingly dependent on foreign trade and military expertise to help maintain their positions and local pre-eminence. The Battle of Plassey in 1756 is often seen as a historical turning point at which a British force defeated the Nawab (aristocratic ruler) of Bengal, yet the importance of the battle can be overestimated and needs to be seen in the context of long-term weakening of Indian economical and military strength. The Maratha Empire succumbed to British rule in 1803, leaving only

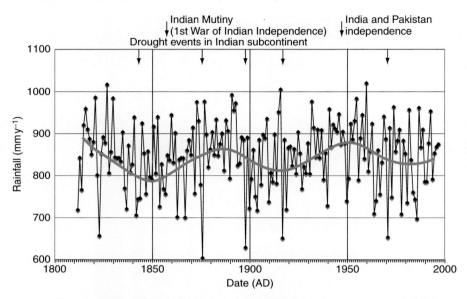

Figure 6.15 The 20 y running average of the Indian summer monsoon rainfall for the last 185 y, showing prominent ∼60 y long-term cycles (data collected by Indian Institute of Tropical Meteorology (IITM), published by Agnihotri *et al.*, 2002). Arrows highlight years of especially low rainfall that correlate to known times of drought and famine in Bengal and other parts of the sub-continent.

the river-irrigated Punjab outside their sphere of influence in the early nineteenth century.

No doubt the strengthening of the summer monsoon into the early nineteenth century generated profits from plantations and helped the establishment of British commercial and colonial power in India, yet by the mid-1830s rainfall had started to fall again (Figure 6.15) and in 1842 the country suffered a severe drought-induced famine. The 1857 revolt against the British, known as the Indian Mutiny or the First War of Indian Independence, followed a protracted period of lower than average rainfall, with strong drought in 1855 (Sontakke *et al.*, 1993). The hardships endured by the local people in these harsh economic circumstances, coupled with the weakened ability of the colonial government to respond to social and political stresses would have played an important role in making the mutiny not only possible but likely. In contrast, although there was a strong drought in 1876–1878 this was at a time of generally higher monsoon rainfalls and resulted in much less social disruption. Droughts and famine induced by failing monsoon rains in 1896–1905 and 1918 (Figure 6.15) came against the backdrop of a generally weaker summer monsoon.

It is noteworthy that the first few decades of the twentieth century were times of lower than average rainfall and were also a time when the movement for Indian Independence was starting to be become stronger, culminating in independence and partition of India and Pakistan in 1947. Although a number of key political events, together with the stresses induced by the First and Second World Wars, were pivotal in undermining British rule on the subcontinent it would be misleading to underestimate the importance of climate-triggered economic stresses in weakening the colonial government there. Crop failures and drought directly influenced the economic state of the general population and made them receptive to the calls for independence, so effectively harnessed by the Congress Party and its allies.

6.6   Monsoon and religion

Monsoon rains have been instrumental to the economic and technical development of human civilizations in Asia, but they have also been important to the spiritual life of people in the region. Praying to the gods for rain to sustain crops and livestock is not unique to Asia, yet it is only in Asia that the rains and rivers themselves have achieved immortal status. This is especially true in South Asia where Hindus consider the monsoon-fed Ganges River to be holy and a place of pilgrimage, as well as a place where the devout spread the ashes of their dead. The cremation of a dead body at the banks of the Ganges and throwing the remains into its water, even when the bodies are burnt elsewhere, is thought to be propitious. It is preferred that the bones of the deceased are brought to the Ganges and cast into the holy river, because it is believed that this leads to salvation of the deceased. Hindus particularly choose this river for holy rituals because the merits of works performed here are believed to be manifold in their results compared to the same acts performed elsewhere.

6.6.1   *Holy monsoon rivers*

The Ganges River is personified by the Goddess Ganga, who is usually represented as a beautiful woman, often depicted with a fish's tail in place of legs, and riding on the Makara, a crocodile-like water monster (Figure 6.16(a)). There are several legends that give various versions of the birth of Ganga. According to one version, the sacred water in the creator god's (Brahma) water vessel became personified as a maiden, Ganga. According to another legend, Brahma had reverently washed the feet of Vishnu, the sustainer god, and collected this water in his vessel from which Ganga then emerged. According to yet a third version, Ganga was the daughter of Himavan, king of the mountains, and his consort Mena.

Bhagiratha, a great king in Hindu mythology, prayed to Brahma that Ganga should come down to Earth so that the souls of Bhagiratha's ancestors would be

Figure 6.16 (a) Bronze statue of the Goddess Ganga, sitting on her Makara, a crocodile-like animal. She holds a water lily in one right hand and in one left hand a lute.
(b) Image of the Goddess Sarasvati holding in two of her four hands a vina instrument, an akshamala (prayer beads) in the other right hand, and a pustaka (book) in the other left, which represents the knowledge of all sciences.

able to go to heaven. Brahma agreed and ordered Ganga to go down to Earth and then on to the nether regions. However, Ganga felt that this was insulting and decided to sweep the whole earth away as she fell from the heavens. In a panic Bhagiratha prayed to the destroyer god, Shiva, that he should break up Ganga's descent, so that when Ganga fell on Shiva's head she became trapped in his hair. Shiva let Ganga out in small streams, so preventing catastrophe. The legend is completed by Ganga traveling to the nether worlds, but at the same time she created a different stream to remain on Earth to help purify unfortunate souls there. It is from this story linking the monsoon-fed river and the greatest gods of the Hindu pantheon that the spiritual importance of the Ganges and the monsoon is derived.

Although Hindus hold the Ganges and its holy cities in special reverence (Figure 6.14(b)) the greatest single demonstration of devotion to the holy waters of the Ganges is during the Kumbh Mela, an event that is attended by millions of people on a single day held four times every twelve years. The major event of this festival is a ritual bath at the banks of the rivers in each town. Other activities include singing, mass feeding of holy men, women and the poor, and assemblies where religious doctrines are discussed. At the Kumbh Mela in 2003, held in Nashik, India, a crowd of 70 million persons gathered for the celebrations, making it the largest religious ceremony known on Earth.

Ganga and the Ganges River are central to the faith of Hindus but this is not the only holy river in South Asia. The Saraswati River, mentioned previously as being potentially the primary water source to the Harappan Culture, is also personified as a goddess (Figure 6.16(b)), described in the holy texts of the Rig Veda. The goddess Sarasvati is considered the goddess of learning, as well as of education, intelligence, crafts, arts, and skills. Because she is the consort of Brahma, who is the source of all knowledge, Sarasvati may be treated as knowledge personified. Her name literally means the one who flows, which can be applied to thoughts, words or the flow of a river. She is commonly depicted seated on a lotus holding a stringed instrument, the Vina (Figure 6.16(b)).

6.6.2   Monsoon religious festivals

The Indian calendar consists of a triad: spring, autumn and monsoon, but is also partitioned into three main phases; cold, hot and the rains. The rains cover a period of four months that, under the Sanskrit name *caturmasa*, meaning "the Four Months," reflects the concept of monsoons. Indians use the monsoon to explain various facets of the Hindu religion. Although the arrival of the rains brings relief to farmers there is also an aspect of fear to their onset. The tenseness of the monsoon time is associated with Vishnu's Sleep. Vishnu is believed to retire below the ocean for his four months' sleep. Thus, this is a time when the Earth is deprived of its lord and is at the mercy of demonic powers. Hindu believers in India try to adjust spiritually to the onset of the monsoon through an elaborate cycle of festivals and other observances. All rituals are not necessarily happy ones. For example, because Vishnu, who is the special protector of the newly married, retires on the eleventh day of Asadha in the Hindu calendar, after that day it becomes necessary to postpone any wedding plans for four months. However, Hindu monsoon festivals generally serve as occasions of renewal and revival. The monsoon can be a time of pilgrimage and spiritual renewal. A spectacular example is that of Kamakhya Temple, located on the south bank of the Brahmaputra River in the eastern state of Assam, and which is considered to be one of the most famous sacred places (pithasthans) of India. During the Ambubashi Festival, which occurs during the monsoon period, there is a great rush of pilgrims to Kamakhya from all over the country.

At the end of the monsoon season, Hindus in northern India celebrate the religious holiday of Rama-Lilas, during which there is a reading and re-enactment of the Ramayana, the Hindu holy text in which Sita, the wife Rama Prince of Ayodhya, is abducted by Ravana, a ten-headed demon, who was king of Lanka (modern Sri Lanka). The day is marked by the burning of effigies of Ravana, and fireworks.

6.6.3   *Monsoons and polytheism*

The links between climate and religion go still deeper than influencing the nature and timing of rituals and myths related to Hinduism or Buddhism in Asia. Suzuki (1978) suggested that the monsoonal climate of South and SE Asia itself favored the development of polytheistic religions in this area. He contrasted the monotheism of Christianity, Islam and Judaism, which evolved in the deserts of the Middle East, with the pantheon of gods associated with the religions that evolved under the monsoon. Yasuda (2004) noted that although people tend to focus on the role of the religious leader in the formation of a religion (e.g., Buddha, Mohammed or Jesus), the role of climate and the influence that this has on theology must also be accounted for when trying to understand a given belief system. It is noteworthy that the two belief systems not only evolved under radically different climate regimes but also resulted in quite different views on how humans, the Earth and God interact with and relate to one another. In the Judaeo-Christian tradition God creates life on Earth for the use and profit of mankind under mankind's dominion. In the wilderness of the desert, where life struggles to survive, it would seem logical that a divine being would be responsible for the creation of living things out of nothing and that in due course time and life will end in a final day of judgment.

In contrast, in the forested land that has grown under the influence of summer monsoon rains, life is everywhere and abundant. Tropical forests teem with life and the cycle of birth, life and death are endlessly replayed, resulting in a theology that does not emphasize a beginning or end of creation. Yasuda (2004) and Suzuki (1978) argue that in this monsoon setting human beings form a small part of a much larger web of life, resulting in a deeper respect, even fear of nature, contrasting with the Judaeo-Christian belief in human mastery over nature. The idea that human life and the greater life of the forest might be parts of a greater spiritual entity forms the starting point of animistic religious beliefs, but is also incorporated into parts of Buddhism. The Tendai sect of Buddhism advances the notion that, "All things and everything; mountains, rivers, grasses, trees and the soil of this land can all attain the wisdom of Buddhahood."

Seen in this context, the shift from early polytheistic religions and animism may be seen as a cultural response to drying of the Eurasian climate since ∼8000 y BP This shift is especially noteworthy in the Mediterranean basin, where there was a long-term shift to monotheism, as forests receded. In many polytheistic theologies of the region it is the rain god or storm gods who were the pre-eminent deity. For example Hadad or Ba'al was a very important Semitic storm and rain god, equivalent in name and origin to the Akkadian god Adad. Hadad has been equated with the Anatolian storm-god Teshub, the Egyptian god Set, the Greek god Zeus and the Roman god Jupiter. The importance of rain, and thus the rain god, to prosperity and life in a variety of cultures across the Middle East is clear, and from there the theological leap to a supreme single deity is less challenging than from the more eclectic theology of Hinduism.

Even in South Asia the importance of rain gods is emphasized. In Vedic religion, the Rig Veda texts single out Varuna as the most powerful of the clan of destructive demons. Varuna is lord of the sky, rain and ocean, as well as being a god of law and of the underworld. In the sphere of monsoon influence, storms and rain are seen as both life-givers and destructive forces to be feared. The Hindu Creator God Brahma is himself born through water, via a lotus that grows from the navel of Vishnu at the beginning of the universe. Another legend says that Brahma created himself by first creating water. In this version of the legend he deposited a seed that later became the golden egg from which Brahma was born.

Although the cultural and spiritual response of peoples in Asia to the changing monsoon differs from east to west, it is clear that the sustaining properties of the summer rains have made a large psychological impact on the people and their ideas about their place and role in the world. Gods of the rivers and rains are prominent in the stories and teachings of all religions, with mankind's relative importance in the natural world strongly dependent on the climate.

6.7   Impacts of future monsoon evolution

Scientists have observed that the Asian monsoon has been gaining strength during the past few centuries, possibly triggered by rising global temperatures (Figure 6.17). The Intergovernmental Panel on Climate Change (IPCC) concluded in 1994 that global warming could intensify the monsoon and increase its variability. Global warming driven by greenhouse gases is predicted to cause the Asian summer monsoon to strengthen further in the future (Ashrit et al., 2001; Hu et al., 2000). However, light-scattering aerosols and land clearing may initially reduce monsoon strength. In the longer term, aerosol-control policies and increasing greenhouse-gas emissions are predicted to lead to an intensification of the Asian monsoon, possibly in a rather abrupt fashion (Zickfeld et al., 2005).

As detailed in Chapter 5, marine upwelling records of the foraminifer G. bulloides from the Arabian Sea have shown that there is an established link between the climate of the North Atlantic Ocean and the intensity of the Asian monsoon (Gupta et al., 2003). This in turn suggests that abrupt climate changes could actually weaken the strength of summer monsoon rains in the future.

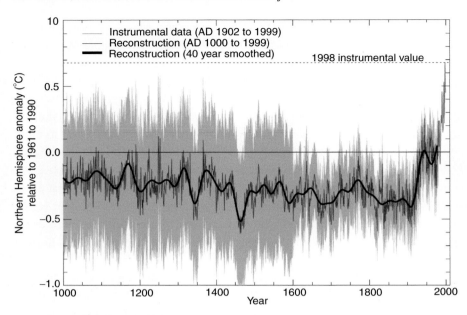

Figure 6.17 Plot of average northern hemispheric temperature anomaly relative to the average values between 1961 and 1990 from the report of the Intergovernmental Panel on Climate Change (2001).

If the North Atlantic should cool quickly as a result of a collapse of the Gulf Stream circulation, triggered by ice melt from Greenland one of the results could be a weakened Asian monsoon and less water for all the people that depend on it. The *G. bulloides* records from the Arabian Sea revealed seven intervals of weakened monsoon since 12 000 y BP coinciding with cold spells in the North Atlantic region. The most intense monsoons occurred at times when the North Atlantic was warmest, that is, during interglacial times. Logically global warming might thus be expected to cause stronger, wetter summer monsoons, and indeed the warming trend of the past ∼200 years is mirrored by heavier annual rains through much of eastern Asia (Figure 6.15), yet this general trend does not preclude dramatic short-term fluctuations.

The monsoon climate system appears to be particularly sensitive to the heat in the North Atlantic, which is controlled not only by global temperatures but also by the Gulf Stream circulation that brings warm water from the Caribbean Sea region to the NE Atlantic Ocean. The circulation of the Gulf Stream is made possible and is balanced by a return flow of cold and saline water, known as the North Atlantic Deep Water (Dickson and Brown, 1994). This conveyor belt of heat transfer maintains the warmth of water in the North Atlantic at much greater values than might otherwise be expected given the high latitudes of the region, but is sensitive to the input of fresh water that can affect the density of the ocean surface water and thus their ability to sink and form the North Atlantic Deep Water (Schmitz and McCartney, 1993). Climate reconstructions show that large influxes of fresh water from melting ice disrupted the heat convection system in the past, resulting in rapid cooling of the North Atlantic (Oppo and Lehman, 1995; Dokken and Jansen, 1999). One such event, 8200 years ago, correlates to a weak monsoon period identified by Gupta et al. (2003).

The Intergovernmental Panel on Climate Change predicts that global temperatures will rise 1–5 °C by the end of the twenty-first century. This could cause large-scale melting of the Greenland ice sheet and result in a surge of freshwater that could cause the North Atlantic circulation to slow or even shut down entirely. If this were to happen, then the intensity of summer monsoon rains throughout Asia would be strongly influenced. Even without the influence of Greenland, ice melting caused by global warming appears to be changing the patterns and strength of rainfall across the North Atlantic region, which in turn must affect the formation of North Atlantic Deep Water. Gradual global warming would be expected to lead to a stronger and more variable monsoon in the twenty-first century as the northern hemisphere continues to warm faster than the tropics. However, once the "tipping point" is reached, where the Greenland ice-cap begins to melt rapidly, then the North Atlantic heat conveyor would abruptly weaken, leading to a sharp and dramatic weakening in the monsoon. Even before this happens, the influence on the large human population of the monsoon region may be significant, as monsoon strength varies rapidly from year to year, bringing drought and flooding in rapid succession.

Changing global climate may affect Asian climate in other ways beyond just controlling the total amount of monsoon rain delivered during the summer season. Changes in the large-scale temperature structure of the northern hemisphere can influence the location of the intertropical convergence zone and the location of monsoon weather fronts. Typhoons accompany the summer rains and their intensity and trajectories are important to people living in their paths in eastern Asia because of their unrivaled ability to cause widespread damage to crops, as well as urban centers close to the coast. Historical records in China include mention of the first documented typhoon landfall in 816 AD (Louie and Liu, 2003). Liu et al. (2001) used historical records to reconstruct the number of major typhoons striking the Asian mainland over the last 1000 years, as well as the location of their landfalls (Figure 6.18). These researchers showed that the two periods of most frequent typhoon strikes in Guangdong province, southern China, (1660–1689 and 1850–1880) coincide with the times of the two coldest and driest periods in northern and central Asia during the Little Ice Age. The predominant storm tracks appear to have shifted to the south during these cold

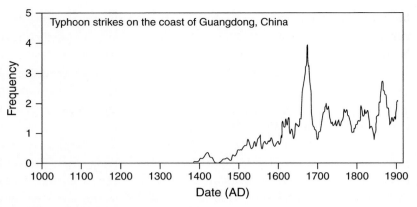

Figure 6.18 Diagram showing the variations at which typhoons have struck the coast of Guangdong province, southern China since 1000 AD, modified from Liu *et al.* (2001). The increase mirrors the general trend to global warming seen in ice-core records over this timespan.

periods, resulting in fewer landfalls in Japan and eastern China but more typhoons in southern China.

Spectral analysis revealed a 50 y cycle in typhoon landfall frequency that Liu *et al.* (2001) correlated with the North Pacific Index for climate variability. However, what is most worrying for people now living in Guangdong, part of the industrial heartland of the modern Chinese economy, is that the number of typhoons striking the coast has risen with increasing global temperatures over the past 500 years. A collapse of the Gulf Stream cooling the North Atlantic region could make matters worse again for Guangdong if cooling in that area drove further cooling in Central Asia, forcing the storm track to stay to the south and simultaneously intensify.

Changes in monsoon intensity may also have serious environmental consequences in regions not directly influenced by this system, as a result of the teleconnections between different climatic systems. Indonesia's climate is known to vary significantly from year to year as a result of the El Niño Southern Oscillation (ENSO) system. Drier and wetter conditions are associated with changes in sea-surface temperature and atmospheric pressure across the tropical Pacific (Overpeck and Cole, 2007), but there is also influence from the Indian Ocean Dipole. This latter system is a cyclical system of changing sea-surface temperatures and salinities across the equatorial Indian Ocean driven by strong cross-equatorial winds. Geochemical records derived from corals from Indonesia have shown that when the summer monsoon was stronger in the Middle Holocene the dipole was stronger and the resultant droughts in Indonesia were of longer duration (Abram *et al.*, 2007). Future changes in monsoon intensity thus have the potential to cause significant socio-economic stress both in Asia and far beyond, even in those regions where other climate systems might be expected to be dominant.

6.8   Summary

High resolution climate records from caves, tree rings, ice cores and lake deposits show significant variation in monsoon strength since 8 ka, with a general trend toward a weaker summer monsoon across Asia after an Early Holocene maximum. Drying of the continent correlates with the general increase in cultural sophistication and it is suggested that increasing environmental stress was an important trigger for the start of agriculture and the formation of the first urban cultures. Some of the earliest civilizations are known from the Indus (Harappan) and Yellow River (Qijia and Longshan) valleys, developing along with those in Mesopotamia and Egypt. These cultures collapsed around 4200 y BP at a time of rapid monsoon weakening, owing to direct negative impact on regional agriculture and more indirectly through changes in the river systems.

Climatic variations since ~2200 BCE appear to have played an important part in allowing populations to prosper or alternatively causing crisis and societal disintegration. Changes in several Chinese dynasties (Shang, Qin and Tang), as well as the Mogul Empire in India correlate with periods of weakening summer monsoon rains. Future climate changes driven by global warming are predicted to cause a stronger summer monsoon, yet in the shorter term ice-cap melting in the North Atlantic could result in a weakened monsoon, with possibly disastrous consequences for the billions of people now living in the influence of the monsoon. Teleconnections to other climate systems, including the Indian Ocean Dipole and ENSO suggest that changes in the monsoon may also cause environmental stress outside core monsoon regions, with more intense and longer droughts predicted in much of Indonesia. A more detailed understanding of monsoon behavior is crucial not only for science, but for all those interested in the economic, military and cultural state of Asia.

References

Abe, M., Kitoh, A. and Yasunari, T. (2003) An evolution of the Asian summer monsoon associated with mountain uplift – simulation with the MRI atmosphere-ocean coupled GCM. *J. Meteor. Soc. Japan*, **81**, 909–933.

Abram, N. J., Gagan, M. K., Liu, Z., Hantoro, W. S., McCulloch, M. T. and Suwargadi, B. W. (2007) Seasonal characteristics of the Indian Ocean Dipole during the Holocene epoch. *Nature*, **445**, 299–302.

Agnihotri, R., Dutta, K., Bhushan, R. and Somayajulu, B. L. K. (2002) Evidence for solar forcing on the Indian monsoon during the last millennium. *Earth Planet. Sci. Lett.*, **198**, 521–527.

Aitchison, J. C., Ali, J. R. and Davis, A. M. (2007) When and where did India and Asia collide? *J. Geophys. Res.*, **112**, B05423, doi: 10.1029/2006JB004706.

Alam, M. and Subrahmanyam, S. (1998) *The Mughal state, 1526–1750*. Delhi, India: Oxford University Press, p. 455.

Albright, W. F. (1944) Ur excavations, vol. V: the Ziggurat and its surroundings. *American J. Archaeo.*, **48**, 303–305.

Ali, J. R. and Aitchison, J. C. (2005) Greater India. *Earth Sci. Rev.*, **72**, 169–188.

Alley, R. B., Meese, D. A., Shuman, C. A. *et al.* (1993) Abrupt increase in Greenland snow accumulation at the end of the Younger Dryas event. *Nature*, **362**, 527–529.

Altabet, M. A., Francois, R., Murray, D. W. and Prell, W. L. (1995) Climate-related variations in denitrification in the Arabian Sea from sediment $^{15}N/^{14}N$ ratios. *Nature*, **373**, 506–509.

Altabet, M. A., Murray, D. W. and Prell, W. L. (1999) Climatically linked oscillations in Arabian Sea denitrification over the past 1 m.y.: implications for the marine N cycle. *Paleoceanography*, **14**, 732–743.

Altabet, M. A., Higginson, M. J. and Murray, D. W. (2002) The effect of millennial-scale changes in Arabian Sea denitrification on atmospheric CO_2. *Nature*, **415**, 159–162.

An, C.-B., Feng, Z. and Tang, L. (2004) Environmental change and cultural response between 8000 and 4000 cal. yr BP in the western Loess Plateau, north-west China. *J. Quat. Sci.*, **19**, 529–535.

An, C.-B., Tang, L., Barton, L. and Chen, F. H. (2005) Climate change and cultural response around 4000 cal yr B. P. in the western part of Chinese Loess Plateau. *Quat. Res.*, **63**, 347–352.

An, Z. (1991) Radiocarbon dating and the prehistoric archaeology of China. *World Archaeo.*, **23**, 193–200.

An, Z., Kukla, G. J., Porter, S. C. and Xiao, J. (1991) Magnetic susceptibility evidence of monsoon variation on the Loess Plateau of central China during the last 130 000 years. *Quat. Res.*, **36**, 29–36.

An, Z. S., Porter, S. C., Chappell, J. *et al.* (1994) The Luochuan loess sequence over the past 130 ka and records of the Greenland ice cores. *Chinese Sci. Bull.*, **39**, 182–184.

An, Z., Kutzbach, J. E., Prell, W. L. and Porter, S. C. (2001) Evolution of Asian monsoons and phased uplift of the Himalaya–Tibetan plateau since Late Miocene times. *Nature*, **411**, 62–66.

Anand, P., Elderfield, H. and Conte, M. H. (2003) Calibration of Mg/Ca thermometry in planktonic foraminifera from a sediment trap time series. *Paleoceanography*, **18**, 1050, doi: 10.1029/2002PA000846.

Anderson, D. M. and Prell, W. L. (1993) A 300 kyr record of upwelling off Oman during the late Quaternary; evidence of the Asian south-west monsoon. *Paleoceanography*, **8**, 193–208.

Anderson, D. M., Brock, J. C. and Prell, W. L. (1992) Physical upwelling processes, upper ocean environment and the sediment record of the south-west monsoon. In *Upwelling Systems; Evolution Since the Early Miocene*, ed. C. P. Summerhayes, W. L. Prell and K. C. Emeis. London: Geol. Soc. Lond., Spec. Publ., vol. **64**, pp. 121–129.

Anderson, D. M., Overpeck, J. T. and Gupta, A. K. (2002) Increase in the Asian south-west monsoon during the past four centuries. *Science*, **297**, 596–599.

Andersson, J. G. (1923) Essays on the Cenozoic of northern China. *Mem. Geol. Surv. China, Ser. A*, **3**, 1–152.

Ashrit, R. G., Kumar, K. R. and Kumar, K. K. (2001) ENSO-monsoon relationships in a greenhouse warming scenario. *Geophys. Res. Lett.* **28**, 1727–1730.

Audley-Charles, M. G. (2004) Ocean trench blocked and obliterated by Banda forearc collision with Australian proximal continental slope. *Tectonophysics*, **389**, 65–79.

Axelrod, D. I. (1980) Estimating altitudes of Tertiary forests. In *Proceedings of Symposium on Qinghai-Zizang (Tibet) Plateau*, Beijing, China: Academica Sinica, 0–2.

Barber, D. C., Dyke, A., Hillaire, M. C. *et al.* (1999) Forcing of the cold event of 8 200 years ago by catastrophic drainage of Laurentide lakes. *Nature*, **400**, 344–348.

Bard, E., Hamelin, B., Arnold, M. *et al.* (1996) Deglacial sea-level record from Tahiti corals and the timing of global meltwater discharge. *Nature*, **382**, 241–244.

Bassinot, F. C., Labeyrie, L. D., Vincent, E. *et al.* (1994) The astronomical theory of climate and the age of the Brunhes–Matuyama magnetic reversal. *Earth Planet. Sci. Lett.*, **126**, 91–108.

Beaumont, C., Jamieson, R. A., Nguyen, M. H. and Medvedev, S. (2004) Crustal channel flows: 1. Numerical models with applications to the tectonics of the Himalayan–Tibetan orogen. *J. Geophys. Res.* **109**, B06406, doi: 10.1029/2003JB002809.

Béchennec, F., Le Métour, J., Platel, J. P. and Roger, J. (1993) *Explanatory notes to the geological map of the Sultanate of Oman*. Muscat, Oman: Directorate General of Minerals, Oman Ministry of Petroleum and Minerals, p. 93.

Bettinger, R. L., Barton, L., Richerson, P. J. *et al.* (2007) The transition to agriculture in North-western China. In *Late Quaternary Climate Change and Human Adaptation in Arid China*, ed. D. B. Madsen, F. Chen and X. Gao, Amsterdam: Elsevier, Develop. Quat. Sci., vol. **9**, pp. 83–101.

Bhattacharya, A. (1989) Vegetation and climate during the last 30 000 years in Ladakh. *Palaeogeog., Palaeoclimat., Palaeoeco.*, **73**, 25–38.

Bird, M. I. and Cali, J. A. (1998) A million-year record of fire in sub-Saharan Africa. *Nature*, **394**, 767–769.

Bird, P. (1978) Initiation of intracontinental subduction in the Himalaya. *J. Geophys. Res.*, **83**, 4975–4987.

Biscaye, P. E. (1965) Mineralogy and sedimentation of Recent deep-sea clay in the Atlantic Ocean and adjacent seas and oceans. *Geol. Soc. Amer. Bull.*, **76**, 803–831.

Blisniuk, P. M., Hacker, B. R., Glodny, J. *et al.* (2001) Normal faulting in central Tibet since at least 13.5 Myr ago. *Nature*, **412**, 628–632.

Bohren, C. F. and Albrecht, B. A. (1998) *Atmospheric Thermodynamics*, Oxford: Oxford University Press, p. 404.

Bond, G. C., Heinrich, H., Broecker, W. S. *et al.* (1992) Evidence for massive discharges of icebergs into the North Atlantic ocean during the last glacial period. *Nature*, **360**, 245–249.

Bond, G., Showers, W., Cheseby, M. *et al.* (1997) A pervasive millennial-scale cycle in North Atlantic Holocene and glacial climates. *Science*, **278**, 1257–1266.

Bond, G., Kromer, B., Beer, J. *et al.* (2001) Persistent solar influence on North Atlantic Climate during the Holocene. *Science*, **294**, 2130–2136.

Bookhagen, B. and Burbank, D. W. (2006) Topography, relief and TRMM-derived rainfall variations along the Himalaya. *Geophys. Res. Lett.*, **33**, L08405, doi: 10.1029/2006GL026037.

Bookhagen, B., Thiede, R. C. and Strecker, M. R. (2005a) Late Quaternary intensified monsoon phases control landscape evolution in the north-west Himalaya. *Geology*, **33**, 149–152.

Bookhagen, B., Thiede, R. C. and Strecker, M. R. (2005b) Abnormal monsoon years and their control on erosion and sediment flux in the high, arid north-west Himalaya. *Earth Planet. Sci. Lett.*, **231**, 131–146.

Brass, G. W. and Raman, C. V. (1991) Clay mineralogy of sediments from the Bengal Fan. *Proc. Ocean Drill. Prog., Sci. Res.*, ed. J. R. Cochran, D. A. V. Stow *et al.*, 116, College Station, TX: Ocean Drilling Program, pp. 35–42.

Bray, H. E. and Stokes, S. (2004) Temporal patterns of arid-humid transitions in the south-eastern Arabian Peninsula based on optical dating. *Geomorph.*, **59**, 271–280.

Broecker, W. S. (1994) Massive iceberg discharges as triggers for global climate change. *Nature*, **372**, 421–424.

Brookfield, M. E. (1998) The evolution of the great river systems of southern Asia during the Cenozoic India–Asia collision; rivers draining southwards. *Geomorph.*, **22**, 285–312.

Bryson, R. A. and Swain, A. M. (1981) Holocene variations of monsoon rainfall in Rajasthan. *Quat. Res.*, **16**, 135–145.

Bull, J. M. and Scrutton, R. A. (1990) Fault reactivation in the central Indian Ocean and the rheology of oceanic lithosphere. *Nature*, **344**, 855–858.

Bull, J. M. and Scrutton, R. A. (1992) Seismic reflection images of intraplate deformation, central Indian Ocean and their tectonic significance. *J. Geol. Soc.*, **149**, 955–966.

Burbank, D. W., Derry, L. A. and France-Lanord, C. (1993) Reduced Himalayan sediment production 8 Myr ago despite an intensified monsoon. *Nature*, **364**, 48–50.

Burbank, D. W., Blythe, A. E., Putkonen, J. *et al.* (2003) Decoupling of erosion and precipitation in the Himalayas. *Nature*, **426**, 652–655.

Burchfiel, B. C., Chen, Z., Hodges, K. V. *et al.* (1992) *The South Tibetan Detachment System, Himalayan Orogen: Extension Contemporaneous With and Parallel to Shortening in a Collisional Mountain Belt.* Boulder, CO: Geological Society of America, Geol. Soc. Amer. Spec. Paper, **269**, p. 41.

Burg, J. P., Proust, F., Tapponnier, P. and Chen, G. M. (1983) Deformation phases and tectonic evolution of the Lhasa Block (southern Tibet, China). *Eclog. Geol. Helv.*, **76**, 643–665.

Burns, S. J. and Matter, A. (1995) Geochemistry of carbonate cements in surficial alluvial conglomerates and their paleoclimatic implications, Sultanate of Oman. *J. Sed. Res.*, **65**, 170–177.

Burns, S. J., Matter, A., Frank, N. and Mangini, A. (1998) Speleothem-based paleoclimate record from northern Oman. *Geology*, **26**, 499–502.

Burns, S. J., Fleitmann, D., Mudelsee, M. *et al.* (2002) 780-year annually resolved record of Indian Ocean monsoon precipitation from a speleothem from South Oman. *J. Geophys. Res.*, **107**, 20, 4434.

Burns, S. J., Fleitmann, D., Matter, A., Kramers, J. and Al-Subbary, A. A. (2003) Indian Ocean climate and an absolute chronology over Dansgaard/Oeschger events 9 to 13. *Science*, **301**, 1365–1367.

Bush, A. B. G. (2004) Modeling of late Quaternary climate over Asia: a synthesis. *Boreas*, **33**, 155–163.

Calvache-Archila, J. and Love, C. (2001) Structural provinces in the northern Sohar Basin, offshore Oman. *Abstr. Int. Conf. Geol. Oman*, Muscat, Oman: Oman Ministry of Commerce, p. 27.

Camoin, G. F., Montaggioni, L. F. and Braithwaite, C. J. R. (2004) Late glacial to post glacial sea levels in the western Indian Ocean. *Mar. Geol.*, **206**, 119–146.

Cane, M. A. (1998) A role for the tropical Pacific. *Science*, **282**, 60–61.

Cane, M. A. and Molnar, P. (2001) Closing of the Indonesian seaway as a precursor to East African aridification around 3–4 million years ago. *Nature*, **411**, 157–162.

Cerling, T. E., Wang, Y. and Quade, J. (1993) Expansion of C4 ecosystems as an indicator of global ecological change in the late Miocene. *Nature*, **361** (6410), 344–345.

Cerling, T. E., Harris, J. M., MacFadden, B. J. *et al.* (1997) Global vegetation change through the Miocene/Pliocene boundary. *Nature*, **38**, 153–158.

Chakraborty, A., Nanjundiah, R. S. and Srinivasan, J. (2002) Role of Asian and African orography in Indian summer monsoon. *Geophys. Res. Lett.*, **29**, doi: 10.1029/2002GL015522.

Chen, J., Farrell, J. W., Murray, D. W. and Prell, W. L. (1995) Timescale and paleoceanographic implications of a 3.6 Ma oxygen isotope record from the northeast Indian Ocean (Ocean Drilling Program Site 758). *Paleoceanography*, **10**, 21–48.

Chen, J., Zheng, L., Wiesner, M. G. *et al.* (1998) Estimations of primary production and export production in the South China Sea based on sediment trap experiments. *China Sci. Bull.*, **43**, 583–586.

Chen, J., An, Z. S. and Head, J. (1999) Variation of Rb/Sr ratios in the loess-paleosol sequences of central China during the last 130 000 years and their implications for monsoon paleoclimatology. *Quat. Res.*, **51**, 215–219.

Chen, K. T. and Hiebert, F. T. (1995) The late prehistory of Xinjiang in relation to its neighbors. *J. World Prehist.*, **9**, 243–300.

Chen, M., Wang, R., Yang, L., Han, J. and Lu, J. (2003) Development of East Asian summer monsoon environments in the late Miocene; radiolarian evidence from Site 1143 of ODP Leg 184. *Mar. Geol.*, **201**, 169–177.

Chen, M.-T. and Huang, C. Y. (1998) Ice-volume forcing of winter monsoon climate in the South China Sea. *Paleoceanography*, **13**, 622–633.

Cheng, H., Edwards, R. L., Wang, Y. *et al.* (2006) A penultimate glacial monsoon record from Hulu Cave and two-phase glacial terminations. *Geology*, **34**, 217–220.

Chou, C. and Neelin, J. D. (2001) Mechanisms limiting the southward extent of the South American summer monsoon. *Geophys. Res. Lett.*, **28**, 2433–2436.

Chou, C., Neelin, J. D. and Su, H. (2001) Ocean-atmosphere-land feedbacks in an idealized monsoon. *Quart. J. R. Meteor. Soc.*, **127**, 1869–1891.

Chu, P. C. and Li, R. (2000) South China Sea isopycnal-surface circulation. *J. Phys. Ocean.*, **30**, 2419–2438.

Chung, S. L., Lo, C. H., Lee, T. Y. *et al.* (1998) Diachronous uplift of the Tibet Plateau starting 40 Myr ago. *Nature*, **394**, 769–773.

Clark, C. O., Cole, J. E. and Webster, P. J. (2000) Indian Ocean SST and Indian summer rainfall: predictive relationships and their decadal variability. *J. Clim.*, **13**, 2503–2519.

Clark, M. K. and Royden, L. H. (2000) Topographic ooze; building the eastern margin of Tibet by lower crustal flow. *Geology*, **28**, 703–706.

Clark, M. K., Schoenbohm, L. M., Royden, L. H. *et al.* (2004) Surface uplift, tectonics, and erosion of Eastern Tibet from large-scale drainage patterns. *Tectonics*, **23**, TC1006, doi: 10.1029/2002TC001402.

Clark, M. K., House, M. A., Royden, L. H. *et al.* (2005) Late Cenozoic uplift of south-eastern Tibet. *Geology*, **33**, 525–528.

Clemens, S. C. and Prell, W. L. (1990) Late Pleistocene variability of Arabian Sea summer monsoon winds and continental aridity; Eolian records from the lithogenic component of deep-sea sediments. *Paleoceanography*, **5**, 109–145.

Clemens, S. C. and Prell, W. L. (2003) Data report: preliminary oxygen and carbon isotopes from site 1146, northern South China Sea. *Proc. Ocean Drill. Prog., Sci. Res.*, **184**, 1–8 (online).

Clemens, S. C. and Prell, W. L. (2007) The timing of orbital-scale Indian monsoon changes. *Quat. Sci. Rev.*, **26**, 275–278.

Clemens, S. C., Prell, W., Murray, D., Shimmield, G. and Weedon, G. (1991) Forcing mechanisms of the Indian Ocean monsoon. *Nature*, **353**, 720–725.

Clemens, S. C., Murray, D. W. and Prell, W. L. (1996) Nonstationary phase of the Plio-Pleistocene Asian monsoon. *Science*, **274**, 943–948.

Clift, P. D., Giosan, L., Blusztajn, J. *et al.* (2008) Holocene erosion of the Lesser Himalaya triggered by intensified summer monsoon. *Geology*, **36**, 79–82.

Clift, P. D. (2006) Controls on the erosion of Cenozoic Asia and the flux of clastic sediment to the ocean. *Earth Planet. Sci. Lett.*, **241**, 571–580.

Clift, P. D. and Blusztajn, J. (2005) Re-organization of the western Himalayan river system after five million years ago. *Nature*, **438**, 1001–1003.

Clift, P. D., Lee, J. I., Blusztajn, J. and Clark, M. K. (2002a) Erosional response of South China to arc rifting and monsoonal strengthening recorded in the South China Sea. *Mar. Geol.*, **184**, 207–226.

Clift, P. D., Gaedicke, C., Edwards, R. *et al.* (2002b) The stratigraphic evolution of the Indus Fan and the history of sedimentation in the Arabian Sea. *Mar. Geophys. Res.*, **23**, 223–245.

Clift, P. D., Lee, J. I., Hildebrand, P. *et al.* (2002c) Nd and Pb isotope variability in the Indus River system: implications for sediment provenance and crustal heterogeneity in the Western Himalaya. *Earth Planet. Sci. Lett.*, **200**, 91–106.

Clift, P. D., Layne, G. D. and Blusztajn, J. (2004) The erosional record of Tibetan uplift in the East Asian marginal seas. In *Continent-Ocean Interactions in the East Asian Marginal Seas*, ed. P. D. Clift, P. Wang, D. Hayes, W. Kuhnt. *Amer. Geophys. Union, monogr.*, 149, 255–282.

Clift, P. D., Blusztajn, J. and Nguyen, D. A. (2006a) Large-scale drainage capture and surface uplift in Eastern Tibet before 24 Ma. *Geophys. Res. Lett*, **33**, L19403, doi: 10.1029/2006GL027772.

Clift, P. D., Carter, A., Campbell, I. H. *et al.* (2006b) Thermochronology of mineral grains in the Song Hong and Mekong Rivers, Vietnam. *Geophys., Geochem., Geosyst.*, **7**, Q10005, doi: 10.1029/2006GC001336.

Cline, J. D. and Kaplan, I. R. (1975) Isotopic fractionation of dissolved nitrate during denitrification in the Eastern Tropical North Pacific Ocean. *Mar. Chem.*, **3**, 271–299.

Cochran, J. R. (1990) Himalayan uplift, sea level, and the record of Bengal Fan sedimentation at the ODP leg 116 sites. *Proc. Ocean Drill. Prog., Sci. Res.*, **116**, 397–414.

Coleman, M. and Hodges, K. (1995) Evidence for Tibetan plateau uplift before 14 Myr ago from a new minimum age for east-west extension. *Nature*, **374**, 49–52.

Colin, C., Turpin, L., Bertaux, J., Desprairies, A. and Kissel, C. (1999) Erosional history of the Himalayan and Burman ranges during the last two glacial-interglacial cycles. *Earth Planet. Sci. Lett.*, **171**, 647–660.

Conkright, M., Levitus, S., O'Brien, T. *et al.* (1998) *World Ocean Atlas 1998*, CD-ROM data set documentation, tech. rep. 15. Silver Spring, MD: National Oceanographic Data Center, p. 16.

Conte, M. H. and Weber, J. C. (2002) Plant biomarkers in aerosols record isotopic discrimination of terrestrial photosynthesis. *Nature*, **417**, 639–641.

Cook, E. R., D'Arrigo, R. D. and Briffa, K. R. (1998) A reconstruction of the North Atlantic oscillation using tree-ring chronologies from North America and Europe. *Holocene*, **8**, 9–17.

Cook, E. R., Woodhouse, C., Eakin, C. M., Meko, D. M. and Stahle, D. W. (2004) Long-term aridity changes in the western United States. *Science*, **306**, 1015–1018.

Copeland, P., Harrison, T. M., Kidd, W. S. F., Xu R. and Zhang Y. (1987) Rapid Miocene acceleration of uplift in the Gangdese belt, Xizang (southern Tibet), and its bearing on accommodation mechanisms of the India–Asia collision. *Earth Planet. Sci. Letts.*, **86**, 240–252.

Cullen, H. M., deMenocal, P. B., Hemming, S. *et al.* (2000) Climate change and the collapse of the Akkadian empire: evidence from the deep sea. *Geology*, **28**, 379–382.

Curray, J. R. (1994) Sediment volume and mass beneath the Bay of Bengal. *Earth Planet. Sci. Lett.*, **125**, 371–383.

Curray, J. R. and Moore, D. G. (1971) Growth of the Bengal deep-sea fan and denudation in the Himalayas. *Geol. Soc. Amer. Bull.*, **82**, 563–572.

Curray, J. R., Emmel, F. J. and Moore, D. G. (2003) The Bengal Fan: morphology, geometry, stratigraphy, history and processes. *Mar. Petrol. Geol.*, **19**, 1191–1223.

Currie, B. S., Rowley, D. B. and Tabor, N. J. (2005) Middle Miocene paleoaltimetry of southern Tibet; implications for the role of mantle thickening and delamination in the Himalayan Orogen. *Geology*, **33**, 181–184.

Curry, W. B., Ostermann, D. R., Guptha, M. V. S. and Itekkot, V. (1992) Foraminiferal production and monsoonal upwelling in the Arabian Sea; evidence from sediment traps. In *Upwelling Systems; Evolution Since the Early Miocene*, ed. C. P. Summerhayes, W. L. Prell and K. C. Emeis, *Geol. Soc. Lond., Spec. Publ.*, 64, 93–106.

Dadson, S., Hovius, N. Chen, H. *et al.* (2003) Links between erosion, runoff variability and seismicity in the Taiwan orogen. *Nature*, **426**, 648–651.

Dahl, K. A. and Oppo, D. W. (2006) Sea-surface temperature pattern reconstructions in the Arabian Sea. *Paleoceanography*, **21**, PA1014, doi: 10.1029/2005PA001162.

Dannenmann, S., Linsley, B. K., Oppo, D. W., Rosenthal, Y. and Beaufort, L. (2003) East Asian monsoon forcing of suborbital variability in the Sulu Sea during Marine Isotope Stage 3; link to Northern Hemisphere climate. *Geochem., Geophys., Geosyst.*, **4**, 1, doi: 10.1029/2002GC000390.

Dansgaard, W., Johnsen, S. J., Clausen, H. B. *et al.* (1993) Evidence for general instability of past climate from a 250-kyr ice-core record. *Nature*, **364**, 218–220.

Davies, T. A., Kidd, R. B. and Ramsay, A. T. S. (1995) A time-slice approach to the history of Cenozoic sedimentation in the Indian Ocean. *Sed. Geol.*, **96**, 157–179, 1995.

DeCelles, P. G., Robinson, D. M., Quade, J. *et al.* (2001) Stratigraphy, structure and tectonic evolution of the Himalayan fold–thrust belt in western Nepal. *Tectonics* **20**, 487–509.

deMenocal, P. B. (2001) Cultural responses to climate change during the late Holocene. *Science*, **292**, 667–673.

deMenocal, P. B., Ortiz, J., Guilderson, T. and Sarnthein, M. (2000) Coherent high- and low-latitude climate variability during the Holocene warm period. *Science*, **288**, 2198–2202.

Dercourt, J., Ricou, L. E. and Vrielinck, B. eds., (1993) *Atlas Tethys Palaeoenvironmental Maps*, Paris: Gauthier-Villars, pp. 1–307.

Derry L. A. and France-Lanord C. (1996) Neogene Himalayan weathering history and river ^{87}Sr/^{86}Sr: impact on the marine Sr record. *Earth Planet. Sci. Lett.*, **142**, 59–76.

Derry, L. A. and France-Lanord, C. (1997) Himalayan weathering and erosion fluxes; climate and tectonic controls. In *Tectonic Uplift and Climate Change*, ed. W. F. Ruddiman, New York: Plenum Press, pp. 289–312.

Dettman, D. L., Kohn, M. J., Quade, J. *et al.* (2001) Seasonal stable isotope evidence for a strong Asian monsoon throughout the past 10.7 m.y. *Geology*, **29**, 31–34.

Dettman, D. L., Fang, X., Garzione, C. N. and Li, J. (2003) Uplift-driven climate change at 12 Ma; a long δ^{18}O record from the NE margin of the Tibetan Plateau. *Earth Planet. Sci. Lett.*, **214**, 267–277.

Dickson, R. R. and Brown, J. (1994) The production of North Atlantic Deep Water: sources, rates, and pathways. *J. Geophys. Res.*, **9**, 12 319–12 342.

Ding, L., Kapp, P., Zhong, D. and Deng, W. (2003) Cenozoic volcanism in Tibet; evidence for a transition from oceanic to continental subduction. *J. Petrol.*, **44**, 1833–1865.

Ding, Z., Liu, T., Rutter, N. W. *et al.* (1995) Ice-volume forcing of East Asian winter monsoon variations in the past 800 000 years. *Quat. Res.*, **44**(2), 149–159.

Ding, Z. L., Sun, J. M., Liu, T. S. *et al.* (1998) Wind-blown origin of the Pliocene red clay formation in the central Loess Plateau, China. *Earth Planet. Sci. Lett.*, **161**(1–4), 135–143.

Ding, Z. L., Xiong, S. F., Sun, J. M. *et al.* (1999) Pedostratigraphy and paleomagnetism of an approximately 7.0 Ma Eolian loess–red clay sequence at Lingtai, Loess Plateau, north-central China and the implications for paleomonsoon evolution. *Palaeogeog., Palaeoclim., Palaeoeco.*, **152**, 49–66.

Ding, Z. L., Sun, J. M., Yang, S. L. and Liu, T. S. (2001) Geochemistry of the Pliocene red clay formation in the Chinese Loess Plateau and implications for its origin, source provenance and paleoclimate change. *Geochim. Cosmochim. Acta*, **65**, 901–913.

Dokken, T. M. and Jansen, E. (1999) Rapid changes in the mechanism of ocean convection during the last glacial period. *Nature*, **401**, 458–461.

Doose-Rolinski, H., Rogalla, U., Scheeder, G., Lückge, A. and von Rad, U. (2001) High-resolution temperature and evaporation changes during the late Holocene in the north-eastern Arabian Sea. *Paleoceanography*, **16**(4), 358–367.

Duplessy, J. C. (1982) Glacial to interglacial contrasts in the northern Indian Ocean. *Nature*, **295**, 494–498.

Edwards, R. L., Chen, J. H. and Wasserburg, G. J. (1987) ^{238}U-^{234}U-^{230}-Th-^{232}Th systematics and the precise measurement of time over the past 500 000 years. *Earth Planet. Sci. Lett.*, **81**, 175–192.

Elston, R. G., Xu, C., Madsen, D. B. *et al.* (1997) New dates for the North China Mesolithic. *Antiquity*, **71**, 985–993.

Emanuel, K. A. (1986) An air–sea interaction theory for tropical cyclones. Part I: steady state maintenance. *J. Atmos. Sci.*, **43**, 585–604.

Emanuel, K. A. (2000) A statistical analysis of hurricane intensity. *Mon. Wea. Rev.*, **128**, 1139–1152.

England, P. C. and Houseman, G. (1989) Extension during continental convergence with application to the Tibetan Plateau. *J. Geophys. Res.*, **94**, 17561–17579.

England, P. C. and Searle, M. P. (1986) The Cretaceous–Tertiary deformation of the Lhasa Block and its implications for crustal thickening in Tibet. *Tectonics*, **5**, 1–14.

Enzel, Y., Ely, L. L., Mishra, S. *et al.* (1999) High-resolution Holocene environmental changes in the Thar Desert, north-western India. *Science*, **284**, 125–128.

Esper, J., Cook, E. R. and Schweingruber, F. H. (2002) Low-frequency signals in long chronologies for reconstructing past temperature variability. *Science*, **295**, 2250–2253.

Esper, J., Frank, D. C., Wilson, R. J. S., Büntgen, U. and Treydte, K. (2007) Uniform growth trends among central Asian low and high elevation juniper tree sites. *Trees*, **21**, 141–150.

Fairbanks, R. G. (1989) A 17 000-year glacio-eustatic sea level record: influence of glacial melting rates on Younger Dryas event and deep-ocean circulation. *Nature*, **342**, 637–642.

Fang, J. Q. (1991) Lake evolution during the past 30 000 years in China, and its implications for environmental change. *Quat. Res.*, **36**, 37–60.

Fang, X., Ono, Y., Fukusawa, H. *et al.* (1999) Asian summer monsoon instability during the past 60 000 years; magnetic susceptibility and pedogenic evidence from the western Chinese Loess Plateau. *Earth Planet. Sci. Lett.*, **168**, 219–232.

Fang, X., Garzione, C., Van der Voo, R., Li, J. and Fan, M. (2003) Flexural subsidence by 29 Ma on the NE edge of Tibet from the magnetostratigraphy of Linxia Basin, China. *Earth Planet. Sci. Lett.*, **210**, 545–560.

Farley, K. A. (2000) Helium diffusion from apatite; general behavior as illustrated by Durango fluorapatite. *J. Geophys. Res.*, **105**, 2903–2914.

Fasullo, J. and Webster, P. J. (2003) A hydrological definition of the Indian summer monsoon onset and withdrawal. *J. Clim.*, **16**, 3200–3211.

Feng, X., Cui, H., Tang, K. and Conkey, L. E. (1999) Tree-ring δD as an indicator of Asian monsoon intensity. *Quat. Res.*, **51**, 262–266.

Feng, Z.-D., An, C. B., Tang, L. Y. and Jull, A. J. T. (2004) Stratigraphic evidence of a Megahumid climate between 10 000 and 4000 years BP in the western part of the Chinese Loess Plateau. *Global Planet. Change*, **43**, 145–155.

Fielding, E. J. (1996) Tibet uplift and erosion. *Tectonophysics*, **260**, 55–84.

Findlater, J. (1969) Interhemispheric transport of air in the lower troposphere over the western Indian Ocean. *Quart. J. R. Meteor. Soc.*, **95**, 400–403.

Fine, R. A., Lukas, R., Bingham, F. M., Warner, M. J. and Gammon, R. H. (1994) The Western Equatorial Pacific: a water mass crossroads. *J. Geophys. Res.*, **99**, 25 063–25 080.

Fleitmann, D., Burns, S. J., Neff, U. *et al.* (2003) Holocene forcing of the Indian monsoon recorded in a stalagmite from southern Oman. *Science*, **300**, 1737–1739.

Fleitmann, D., Burns, S. J., Neff, U. *et al.* (2004) Palaeoclimatic interpretation of high-resolution oxygen isotope profiles derived from annually laminated speleothems from southern Oman. *Quat. Sci. Rev.*, **23**, 935–945.

Fluteau, F., Ramstein, G. and Besse, J. (1999) Simulating the evolution of the Asian and African monsoons during the past 30 Myr using an atmospheric general circulation model. *J. Geophys. Res.*, **104**, 11 995–12 018.

France-Lanord, C., Derry, L. and Michard, A. (1993) Evolution of the Himalaya since Miocene time: isotopic and sedimentological evidence from the Bengal Fan. In *Himalayan Tectonics*, ed. P. J. Treloar and M. P. Searle. *Geol. Soc., Lond., Spec. Publ.*, **74**, 603–622.

Frumkin, A. (1991) The Holocene climatic record of the salt caves of Mount Sedom, Israel. *The Holocene*, **1**, 191–200.

Galy, A. and France-Lanord, C. (2001) Higher erosion rates in the Himalaya; geochemical constraints on riverine fluxes. *Geology*, **29**, 23–26.

Garzione, C. N., Quade, J., DeCelles, P. G. and English, N. B. (2000) Predicting paleoelevation of Tibet and the Himalaya from δ^{18}O vs. altitude gradients in meteoric water across the Nepal Himalaya. *Earth Planet. Sci. Lett.*, **183**, 215–229.

Garzione, C. N., Dettman, D. L. and Horton, B. K. (2004) Carbonate oxygen isotope paleoaltimetry; evaluating the effect of diagenesis on paleoelevation estimates for the Tibetan Plateau. *Palaeogeog., Palaeoclim., Palaeoeco.*, **212**, 119–140.

Garzione, C. N., Ikari, M. J. and Basu, A. R. (2005) Source of Oligocene to Pliocene sedimentary rocks in the Linxia basin in north-eastern Tibet from Nd isotopes: implications for tectonic forcing of climate. *Geol. Soc. Amer. Bull.*, **117**, 1156–1166.

Gasse, F., Arnold, M., Fontes, J. C. *et al.* (1991) A 13 000-year climate record from western Tibet. *Nature*, **353**, 742–745.

George, A. D., Marshallsea, S. J., Wyrwoll, K. H., Jie, C. and Yanchou, A. (2001) Miocene cooling in the northern Qilian Shan, north-eastern margin of the Tibetan Plateau, revealed by apatite fission-track and vitrinite-reflectance analysis. *Geology*, **29**, 939–942.

Ghose, B., Kar, A. and Husain, Z. (1979) The lost courses of the Saraswati River in the Great Indian Desert; new evidence from Landsat imagery. *Geograph. J.*, **145**, 446–451.

Giosan, L., Flood, R. D., Grutzner, J. and Mudie, P. (2002) Paleoceanographic significance of sediment color on western North Atlantic Drifts: II. Late Pliocene-Pleistocene sedimentation. *Mar. Geol.*, **189**, 43–61.

Giosan, L., Clift, P. D., Blusztajn, J. *et al.* (2006) On the control of climate- and human-modulated fluvial sediment delivery on river delta development: the Indus. *Eos Trans.*, **87**(52), OS14A–04.

Godfrey, J. S. (1996) The effect of the Indonesian throughflow on ocean circulation and heat exchange with the atmosphere: a review. *J. Geophys. Res.*, **101**, 12 217–12 238.

Godfrey, J. S. and Golding, T. J. (1981) The Sverdrup relation in the Indian Ocean, and the effect of Pacific-Indian Ocean throughflow on Indian Ocean circulation and on the East Australian Current. *J. Phys. Ocean.*, **11**, 771–779.

Godfrey, J. S., Alexiou, A., Ilahude, A. G. *et al.* (1995) *The Role of the Indian Ocean in the Global Climate System: Recommendations Regarding the Global Ocean Observing System.* Report of the Ocean Observing System Development Panel, College Station, TX, USA: Texas A & M University, p. 89.

Godin, L., Grujic, D., Law, R. D. and Searle, M. P. (2006) Channel flow, ductile extrusion and exhumation in continental collision zones: an introduction. In *Channel Flow, Ductile Extrusion and Exhumation in Continental Collision Zones,* ed. Law, R. D., Searle, M. P. and Godin, L. *Geol. Soc. Lond. Spec. Publ.,* **268**, 1–24.

Gong, Z., Chen, H., Wang, Z., Cai, F. and Luo, G. (1987) The epigenetic geochemical types of loess in China. In *Aspects of Loess Research,* ed. T. Liu, Beijing: China Ocean Press, pp. 135–150.

Goodbred, S. L. and Kuehl, S. A. (1999) Holocene and modern sediment budgets for the Ganges–Brahmaputra river system; evidence for highstand dispersal to flood-plain, shelf, and deep-sea depocenters. *Geology,* **27**, 559–562.

Goodbred, S. L. and Kuehl, S. A. (2000) Enormous Ganges–Brahmaputra sediment discharge during strengthened early Holocene monsoon. *Geology,* **28**, 1083–1086.

Gordon, A. L. (2001) Interocean Exchange. In *Ocean Circulation and Climate,* ed. G. Siedler, J. Church and J. Gould, London: Academic Press, pp. 303–314.

Gordon, A. L. and Fine, R. A. (1996) Pathways of water between the Pacific and Indian oceans in the Indonesian seas. *Nature,* **379**, 146–149.

Gordon, A. L., Susanto, R. D. and Vranes, K. (2003) Cool Indonesian throughflow as a consequence of restricted surface layer flow. *Nature,* **425**, 824–828.

Green, P. F., Duddy, I. R., Laslett, G. M. *et al.* (1989) Thermal annealing of fission tracks in apatite: four quantitative modeling techniques and extension to geological timescales. *Chem. Geol.,* **79**, 155–182.

Greenland Ice Sheet Project (1997) *The Greenland Summit ice cores.* (CD-ROM) Boulder, Colorado, University of Colorado, National Snow and Ice Data Center, and Boulder, Colorado, National Geophysical Data Center, World Data Center–A for Paleoclimatology, www.ngdc.noaa.gov/paleo/icecore/greenland/summit/index.html.

Grootes, P. M. and Stuiver, M. (1997) $^{18}O/^{16}O$ variability in Greenland snow and ice with 10^3 to 10^5 yr time resolution. *J. Geophys. Res.,* **102**, 26 455–26 470.

Guillot, S., Hodges, K., Le Fort, P. and Pecher, A. (1994) New constraints on the age of the Manaslu leucogranite: evidence for episodic tectonic denudation in the central Himalayas. *Geology,* **22**, 559–562.

Guo, Z. T., Ruddiman, W. F., Hao, Q. Z. *et al.* (2002) Onset of Asian desertification by 22 Myr ago inferred from loess deposits in China. *Nature,* **416**, 159–165.

Gupta, A. K. (2004) Origin of agriculture and domestication of plants and animals linked to early Holocene climate amelioration. *Current Sci.,* **87**, 54–59.

Gupta, A. K. and Thomas, E. (2003) Initiation of Northern Hemisphere Glaciation and strengthening of the North-east Indian monsoon; Ocean Drilling Program Site 758, eastern equatorial Indian Ocean. *Geology,* **31**, 47–50.

Gupta, A. K., Anderson, D. M. and Overpeck, J. T. (2003) Abrupt changes in the Asian south-west monsoon during the Holocene and their links to the North Atlantic Ocean. *Nature,* **421**, 354–356.

Gupta, A. K., Singh, R. K., Joseph, S. and Thomas, E. (2004) Indian ocean high-productivity event (10–8 Ma): linked to global cooling or to the initiation of the Indian monsoons? *Geology*, **32**, 753–756.

Gupta A. K., Das, M. and Anderson, D. M. (2005) Solar influence on the Indian summer monsoon during the Holocene. *Geophys. Res. Lett.*, **32**, L17703, doi: 10.1029/2005GL022685.

Gupta, S. M. (1999) Radiolarian monsoonal index Pylonüd group responds to astronomical forcing in the last 500 000 years: evidence from the central Indian Ocean. *Man Environ.*, **24**, 99–107.

Gupta, S. P. (1995) *The Lost Sarasvati and the Indus Civilization.* Jodhpur, India: Kusumanjali Prakashan, p. 314.

Hahn, D. G. and Manabe, S. (1975) The role of mountains in the south Asian monsoon circulation. *J. Atmos. Sci.*, **32**, 1515–1541.

Hall, R. (2002) Cenozoic geological and plate tectonic evolution of SE Asia and the SW Pacific; computer-based reconstructions, model and animations. *J. Asian Earth Sci.*, **20**, 353–431.

Haq, B. U., Hardenbol, J. and Vail, P. R. (1987) Chronology of fluctuating sea levels since the Triassic. *Science*, **235**, 1156–1167.

Harada, N., Ahagon, N., Sakamoto, T. *et al.* (2006) Rapid fluctuation of alkenone temperature in the south-western Okhotsk Sea during the past 120 kyr. *Global Planet. Change*, **53**, 29–46.

Harris, N. B. W. (2006) The elevation of the Tibetan Plateau and its impact on the monsoon, *Palaeogeog., Palaeoclim., Palaeoeco.*, **241**, 4–15.

Harrison, T. M., Copeland, P., Kidd, W. S. F. and Yin, A. (1992) Raising Tibet. *Science*, **255**, 1663–1670.

Harrison, T. M., Copeland, P., Hall, S. A. *et al.* (1993) Isotopic preservation of Himalayan/Tibetan uplift, denudation and climatic histories in two molasse deposits. *J. Geology*, **100**, 157–173.

Harrison, T. M., Mahon, K. I., Guillot, S. *et al.* (1995) New constraints on the age of the Manaslu leucogranite; evidence for episodic tectonic denudation in the central Himalaya; discussion and reply. *Geology*, **23**, 476–480.

Harzallah, A. and Sadourny, R. (1997) Observed lead – lag relationships between Indian summer monsoon and some meteorological variables. *Clim. Dyn.*, **13**, 635–648.

Hassan, F. A. (1997) Nile floods and political disorder in early Egypt. In *Third Millennium BCE Climate Change and Old World Collapse*, ed. H. N. Dalfes, G. Kukla, and H. Weiss. New York: Springer, NATO ASI Series **1** (49), pp. 1–24.

Haug, G. H. and Tiedemann, R. (1998) Effect of the formation of the Isthmus of Panama on Atlantic Ocean thermohaline circulation. *Nature*, **393**, 673–676.

Hayden, B. (1981) Research and development in the Stone Age: technological transitions among hunter-gatherers. *Current Anthrop.*, **22**, 519–548.

Hays, J. D., Imbrie, J. and Shackleton, N. J. (1976) Variations in the Earth's orbit; pacemaker of the ice ages. *Science*, **194**, 1121–1132.

Held, I. M. and Hou, A. Y. (1980) Nonlinear axially symmetric circulations in a nearly inviscid atmosphere. *J. Atmos. Sci.*, **37**, 515–533.

Heller, F. and Liu, T. S. (1982) Magnetostratigraphic dating of loess deposits in China. *Nature*, **300**, 431–433.

Hendon, H. H. and Liebmann, B. (1990) A composite study of onset of the Australian summer monsoon, *J. Atmos. Sci.*, **47**, 2227–2240.

Herren, E. (1987) Zanskar shear zone; north-east-south-west extension within the Higher Himalayas (Ladakh, India). *Geology*, **15**, 409–413.

Herzschuh, U. (2006) Palaeo-moisture evolution at the margins of the Asian monsoon during the last 50 ka. *Quat. Sci. Rev.*, **25**, 163–178.

Herzschuh, U. Tarasov, P., Wünnemann, B. and Hartmann, K. (2004) Holocene vegetation and climate of the Alashan Plateau, NW China, reconstructed from pollen data. *Palaeogeog. Palaeoclim., Palaeoeco.*, **211**, 1–17.

Herzschuh, U., Zhang, C., Mischke, S. (2005) A Late Quaternary lake record from the Qilian Mountains (NW China): evolution of the primary production and the water depth reconstructed from macrofossil, pollen, biomarker and isotope data. *Global Planet. Change*, **46**, 361–379.

Hess, S. and Kuhnt, W. (2005) Neogene and Quaternary paleoceanographic changes in the southern South China Sea (Site 1143); the benthic foraminiferal record. *Mar. Micropaleo.*, **54**, 63–87.

Heusser, L. and Morley, J. (1997) Monsoon Fluctuations over the past 350 kyr: high-resolution evidence from north-east Asia / north-west Pacific climate proxies (marine pollen and radiolarians). *Quat. Sci. Rev.*, **16**, 565–581.

Higginson, M. J., Maxwell, J. R. and Altabet, M. A. (2003) Nitrogen isotope and chlorin paleoproductivity records from the northern South China Sea; remote vs. local forcing of millennial- and orbital-scale variability. *Mar. Geol.*, **201**, 223–250.

Higham, C. (1996) *The Bronze Age of South-east Asia*. Cambridge: Cambridge University Press, p. 381.

Hodges, K. V. and Silverberg, D. S. (1988) Thermal evolution of the Greater Himalaya, Garhwal, India. *Tectonics*, **7**, 583–600.

Hodges, K. V., Parrish, R. R. and Searle, M. P. (1996) Tectonic evolution of the central Annapurna Range, Nepalese Himalayas. *Tectonics*, **15**, 1264–1291.

Hodges, K. V., Wobus, C., Ruhl, K., Schildgen, T. and Whipple, K. (2004) Quaternary deformation, river steepening, and heavy precipitation at the front of the Higher Himalayan ranges. *Earth Planet. Sci. Lett.*, **220**, 379–389.

Holbourn, A. E., Kuhnt, W., Simo, J. A. and Li, Q. (2004) Middle Miocene isotope stratigraphy and paleoceanographic evolution of the north-west and south-west Australian margins (Wombat Plateau and Great Australian Bight). *Palaeogeog., Palaeoclim., Palaeoeco.*, **208**, 1–22.

Holton, J. R. (2004) *An Introduction to Dynamic Meteorology*, 3rd edn. San Diego: Academic Press, p. 535.

Hong, Y. T., Wang, Z. G., Jiang, H. B. *et al.* (2001) A 6000-year record of changes in drought and precipitation in north-eastern China based on a delta ^{13}C time series from peat cellulose. *Earth Planet. Sci. Lett.*, **185**, 111–119.

Hong, Y. T., Hong, B., Lin, Q. H. *et al.* (2003) Correlation between Indian Ocean summer monsoon and North Atlantic climate during the Holocene. *Earth Planet. Sci. Lett.*, **211**, 371–380.

Hong, Y. T., Hong, B., Lin, Q. H. *et al.* (2005) Inverse phase oscillations between the East Asian and Indian Ocean summer monsoons during the last 12 000 years and paleo-El Niño. *Earth Planet. Sci. Lett.*, **231**, 337–346.

Hoskins, B. J. and Rodwell, M. J. (1995) A model of the Asian summer monsoon. Part I: the global scale. *J. Atmos. Sci.*, **52**, 1329–1340.

Hostetler, S. W., Clark, P. U., Bartlein, P. J., Mix, A. C. and Pisias, N. J. (1999) Atmospheric transmission of North Atlantic Heinrich events. *J. Geophys. Res.*, **104**, 3947–3952.

Houseman, G. A. and England, P. C. (1993) Crustal thickening versus lateral expulsion in the Indian–Asian continental collision. *J. Geophys. Res.*, **98**, 12 233–12 249.

Houseman, G. A., McKenzie, D. P. and Molnar, P. (1981) Convective instability of a thickened boundary layer and its relevance for the thermal evolution of continental convergent belts. *J. Geophys. Res.*, **86**, 6115–6132.

Hovan, S. A., Rea, D. K., Pisias, N. G. and Shackleton, N. J. (1989) A direct link between the China loess and marine $\delta^{18}O$ records; aeolian flux to the North Pacific. *Nature*, **340**, 296–298.

Hovan, S. A., Rea, D. K. and Pisias, N. G. (1991) Late Pleistocene continental climate and oceanic variability recorded in North-west Pacific sediments. *Paleoceanography*, **6**, 349–370.

Hu, Z.-Z., Latif, M., Roeckner, E. and Bengtsson, L. (2000) Intensified Asian summer monsoon and its variability in a coupled model forced by increasing greenhouse gas concentrations. *Geophys. Res. Lett.*, **27**, 2681–2684.

Huang, C. C., Pang, J., Han, Y. P. and Hou, C. H. (2000) A regional aridity phase and its possible cultural impact during the Holocene Megathermal in the Guanzhong Basin, China. *The Holocene*, **10**, 135–142.

Huang, C. C., Zhao, S., Pang, J. *et al.* (2003) Climatic aridity and the relocations of the Zhou culture in the southern loess plateau of China. *Clim. Change*, **61**, 361–378.

Huang, Q., Cai, B. and Ru, C. (1980) Radiocarbon dating of samples from several salt lakes on the Tibet Plateau and their sedimentary cycles. *Kexue Tongbao*, **25**, 990–995.

Huang, Y., Street-Perrott, F. A., Metcalfe, S. E. *et al.* (2001) Climate change as the dominant control on glacial–interglacial variations in C3 and C4 plant abundance. *Science*, **293**, 1647–1651.

Hughes, M. K., Wu, X., Shao, X. and Garfin, G. M. (1994) A preliminary reconstruction of rainfall in North-Central China since A.D. 1600 from tree-ring density and width. *Quat. Res.*, **42**, 88–99.

Huntington, K. W., Blythe, A. E. and Hodges, K. V. (2006) Climate change and Late Pliocene acceleration of erosion in the Himalaya. *Earth Planet. Sci. Lett.*, **252**, 107–118.

Indermühle, A., Monnin, E., Stauffer, B., Stocker, T. F. and Wahlen, M. (2000) Atmospheric CO_2 concentration from 60 to 20 kyr B.P. from the Taylor Dome ice core, Antarctica. *Geophys. Res. Lett.*, **27**, 735–738.

Ingall, E. and Jahnke, R. (1994) Evidence for enhanced phosphorus regeneration from marine sediments overlain by oxygen depleted waters. *Geochim. Cosmochim. Acta*, **58**, 2571–2575.

IPCC (2001) *Synthesis Report. A Contribution of Working Groups I, II, and III to the Third Assessment Report of the Intergovernmental Panel on Climate Change*, ed. T. R. Watson and Core Writing Team, Cambridge: Cambridge University Press, p. 398.

Irino, T. and Tada, R. (2000) Quantification of aeolian dust (Kosa) contribution to the Japan Sea sediments and its variation during the last 200 ky. *Geochem. J.*, **34**, 59–93.

Irino, T. and Tada, R. (2002) High-resolution reconstruction of variation in aeolian dust (Kosa) deposition at ODP Site 797, the Japan Sea, during the last 200 ka. *Global Planet. Change*, **35**, 143–156.

Janecek, T. R. and Rea, D. K. (1985) Quaternary fluctuations in northern hemispheric tradewinds and westerlies. *Quat. Res.*, **24**, 150–163.

Jarrige, J. F. (1993) Excavations at Mehrgarh: their significance for understanding the background of the Harappan Civilization. In *Harappan Civilization: A Recent Perspective*, ed. G. Possehl, Delhi: Oxford University Press, pp. 125–135.

Jia, G., Peng, P., Zhao, Q. and Jian, Z. (2003) Changes in terrestrial ecosystem since 30 Ma in East Asia: stable isotope evidence from black carbon in the South China Sea. *Geology*, **31**, 1093–1096.

Jin, L. and Su, B. (2000) Natives or immigrants: modern human origin in East Asia. *Nature Rev. Genet.*, **1**, 126–133.

Johnsen, S. J., Clausen, H. B., Dansgaard, W. *et al.* (1992) Irregular glacial interstadials recorded in a new Greenland ice core. *Nature*, **359**, 311–313.

Jones, C. E., Halliday, A. N., Rea, D. K. and Owen, R. M. (1994) Neodymium isotopic variations in North Pacific modern silicate sediment and the insignificance of detrital REE contributions to seawater. *Earth Planet. Sci. Lett.*, **127**, 55–66.

Jung, S. J. A., Davies, G. R., Ganssen, G. and Kroon, D. (2002) Decadal-centennial scale monsoon variations in the Arabian Sea during the early Holocene. *Geochem., Geophys., Geosyst.*, **3** (10), 1060, doi: 10.1029/2002GC000348.

Jung, S. J. A., Davies, G. R., Ganssen, G. M. and Kroon, D. (2004) Stepwise Holocene aridification in NE Africa deduced from dust-borne radiogenic isotope records. *Earth Planet. Sci. Lett.*, **221** (1–4), 27–37.

Kapsner, W. R., Alley, R. B., Shuman, C. A., Anandakrishnan, S. and Grootes, P. M. (1995) Dominant influence of atmospheric circulation in Greenland over the past 18 000 years. *Nature*, **373**, 52–54.

Kar, A. (1984) The Drishadvati River system of India; an assessment and new findings. *Geograph. J.*, **150** (2), 221–229.

Karim, A. and Veizer, J. (2002) Water balance of the Indus River Basin and moisture source in the Karakoram and western Himalayas: implications from hydrogen and oxygen isotopes in river water. *J. Geophys. Res.*, **107** (D18), 4362, doi: 10.1029/2000JD000253.

Kirch, A. (1997) Zur Paläoozeanographie westlich von Luzon (Philippinen). M.Sc. thesis, Kiel, Germany: Kiel University, p. 28.

Kitoh, A. (1997) Mountain uplift and surface temperature changes. *Geophys. Res. Lett.*, **24**, 185–188.

Kitoh, A. (2004) Effect of mountain uplift on East Asian summer climate investigated by a coupled atmosphere–ocean GCM. *J. Clim.*, **17**, 783–802.

Klinck, J. M. and Smith, D. A. (1993) Effect of wind changes during the last glacial maximum on the circulation of the Southern Ocean. *Paleoceanography*, **8**, 427–433.

Koons, P. O., Zeitler, P. K., Chamberlain, C. P., Craw, D. and Meltzer, A. S. (2002) Mechanical links between erosion and metamorphism in Nanga Parbat, Pakistan Himalaya. *Amer. J. Sci.*, **302**, 749–773.

Krishna, K. S., Bull, J. M. and Scrutton, R. A. (2001) Evidence for multiphase folding of the central Indian Ocean lithosphere. *Geology*, **29**, 715–718.

Krissek, L. A. and Clemens, S. C. (1991) Mineralogic variations in a Pleistocene high-resolution Eolian record from the Owen Ridge, western Arabian Sea (Site 722); implications for sediment source conditions and monsoon history. *Proc. Ocean Drill. Prog., Sci. Res.*, **117**, 197–213.

Kroon, D., Steens, T. and Troelstra, S. R. (1991) Onset of Monsoonal related upwelling in the western Arabian Sea as revealed by planktonic foraminifers. *Proc. Ocean Drill. Prog., Sci. Res.*, **117**, 257–263.

Kudrass, H. R., Hofmann, A., Doose, H., Emeis, K. and Erlenkeuser, H. (2001) Modulation and amplification of climatic changes in the Northern Hemisphere by the Indian summer monsoon during the past 80 k.y. *Geology*, **29**, 63–66.

Kuhlemann J. (2001) Post-collisional sediment budget of circum-Alpine basins (Central Europe). *Sci. Mem. Geol., Padova*, **52**, 1–91.

Kuhlemann, J., Frisch, W., Dunkl, I. and Szekely, B. (2001) Quantifying tectonic versus erosive denudation by the sediment budget: the Miocene core complexes of the Alps. *Tectonophysics* **330**, 1–23.

Kuhnt, W., Holbourn, A., Hall, E., Zuvela, M. and Käse, R. (2004) Neogene history of the Indonesian Throughflow. In *Continent–Ocean Interactions Within East Asian Marginal Seas*, ed. P. D. Clift, W. Kuhnt, P. Wang and D. E. Hayes, *Geophys. Monogr. Ser.* **149**, Washington DC: American Geophysical Union, pp. 299–320.

Kukla, G., An, Z. S., Melice, J. L., Gavin, J. and Xiao, J. L. (1990) Magnetic susceptibility record of Chinese loess. *Trans. R. Soc. Edin. Earth Sci.*, **81**, 263–288.

Kutzbach, J. E., Guetter, P. J., Ruddiman, W. F. and Prell, W. L. (1989) Sensitivity of climate to late Cenozoic uplift in southern Asia and the American West; numerical experiments. *J. Geophys. Res.*, **94** (15), 18 393–18 407.

Kutzbach, J. E., Prell, W. L. and Ruddiman, W. F. (1993) Sensitivity of Eurasian climate to surface uplift of the Tibetan Plateau. *J. Geology*, **101**, 177–190.

Kuwae, M., Yoshikawa, S., Tsugeki, N. and Inouchi, Y. (2004) Reconstruction of a climate record for the past 140 kyr based on diatom valve flux data from Lake Biwa, Japan. *J. Paleolim.*, **32**, 19–39.

Lal, B. B. (1997) *The Earliest Civilization of South Asia*. Delhi: Aryan Books International, p. 308.

Li, C., Chen, Q., Zhang, J., Yang, S. and Fan, D. (2000) Stratigraphy and paleoenvironmental changes in the Yangtze delta during late Pleistocene. *J. Asian Earth Sci.*, **18**, 453–469.

Li, S., Zheng, B. and Jiao, K. (1989) Preliminary research on lacustrine deposits and lake evolution on the southern slope of the west Kunlun Mountains. *Bull. Glacier Res.*, **7**, 169–176.

Li, X. H., Wei, G., Shao, L. *et al.* (2003) Geochemical and Nd isotopic variations in sediments of the South China Sea: a response to Cenozoic tectonism in SE Asia. *Earth Planet. Sci. Lett.*, **211**, 207–220.

Lindzen, R. S. and Hou, A. Y. (1988) Hadley circulations for zonally averaged heating centered off the equator. *J. Atmos. Sci.*, **45**, 2416–2427.

Liu, K.-B., Shen, C. and Louie, K.-B. (2001) A 1000-year history of typhoon landfalls in Guangdong, Southern China, reconstructed from Chinese historical documentary records. *Ann. Assoc. Amer. Geograph.*, **91** (3), 453–464.

Liu L. (1996) Settlement patterns, chiefdom variability, and the development of early states in North China. *J. Anthrop. Archaeo.*, **15**, 237–288.

Liu, T. S. (1985) *Loess and the Environment.* Beijing: China Ocean Press, p. 251.

Liu, T., Ding, Z. and Rutter, N. (1999) Comparison of Milankovitch periods between continental loess and deep sea records over the last 2.5 Ma. *Quat. Sci. Rev.*, **18**, 1205–1212.

Liu, W., Feng, X., Liu, Y., Zhang, Q. and An, Z. (2004b) $\delta^{18}O$ values of tree rings as a proxy of monsoon precipitation in arid north-west China. *Chem. Geol.*, **206**, 73–80.

Liu, W., Huang, Y., An, Z. *et al.* (2005) Summer monsoon intensity controls C4/C3 plant abundance during the last 35 ka in the Chinese Loess Plateau; carbon isotope evidence from bulk organic matter and individual leaf waxes. *Palaeogeog., Palaeoclim., Palaeoeco.*, **220**, 243–254.

Liu, X. and Yin, Z.-Y. (2002) Sensitivity of East Asian monsoon climate to the uplift of the Tibetan Plateau. *Palaeogeog, Palaeoclimat, Palaeocol.*, **183**, 223–245.

Liu, X. M., Liu, T. S., Xu, T. C., Liu, C. and Chen, M. Y. (1988) The Chinese Loess in Xifeng: I: the primary study on magnetostratigraphy of a loess profile in Xifeng area, Gansu Province. *Geophys. J.*, **93**, 345–348.

Liu, Z. C., Sun, S. Y., Yang, F. and Zhou, Z. H. (1990) Quaternary stratigraphical and chronological studies of Sanhu Region, Qaidam basin (in Chinese). *Sci. China (Ser. B)*, **11**, 1202–1212.

Liu, Z., Trentesaux, A., Clemens, S. C. (2003) Clay mineral assemblages in the northern South China Sea; implications for East Asian monsoon evolution over the past two million years. *Mar. Geol.*, **201**, 133–146.

Liu, Z., Colin, C., Trentesaux, A. *et al.* (2004c) Erosional history of the eastern Tibetan Plateau since 190 kyr ago; clay mineralogical and geochemical investigations from the south-western South China Sea. *Mar. Geol.*, **209**, 1–18.

Liu, Z., Henderson, A. C. G. and Huang, Y. S. (2006) Alkenone-based reconstruction of late Holocene surface temperature and salinity changes in Lake Qinghai, China. *Geophys. Res. Lett.*, **33**, L09707.

Loewe, M. and Shaughness, E. L. (1999) *The Cambridge History of Ancient China: From the Origins of Civilization to 221 BCE.* Cambridge: Cambridge University Press, p. 1180.

Loope, D. B., Rowe, C. M. and Joeckel, R. M. (2001) Annual monsoon rains recorded by Jurassic dunes. *Nature*, **412** (6842), 64–66.

Louie, K. S. and Liu, K.-B. (2003) Earliest historical records of typhoons in China. *J. Hist. Geog.*, **29** (3), 299–316.

Ma, Y., Li, J. and Fan, X. (1998) Pollen-based vegetational and climatic records during 30.6 to 5.0 My from Linxia area, Gansu. *Chinese Sci. Bull.*, **43**, 301–304 (in Chinese).

Madden, R. A. and Julian, P. R. (1972) Description of Global-scale circulation cells in the Tropics with a 40–50 Day Period. *J. Atmos. Sci.*, **29** (6), 1109–1123.

Madden, R. A. and Julian, P. R. (1994) Observations of the 40–50-day Tropical Oscillation – a review. *Mon. Wea. Rev.*, **122**, 814–837.

Madsen, D. B., Li, J., Elston, R. G. *et al.* (1998) The loess/paleosol record and the nature of the Younger Dryas climate in central China. *Geoarchaeo.: Int. J.*, **13** (8), 847–869.

Maher, B. A. (1986) Characterization of soils by mineral magnetic measurements. *Phys. Earth Planet. Interiors*, **42**, 76–92.

Maizels, J. K. and McBean, C. (1990) Cenozoic alluvial fan systems of interior Oman: palaeoenvironmental reconstruction based on discrimination of palaeochannels using remotely sensed data. In *The Geology and Tectonics of the Oman Region*, ed. A. H. F. Robertson, M. P. Searle and A. C. Ries. *Geol. Soc., Lond., Spec. Publ.*, **49**, 565–582.

Mann, M. E., Bradley, R. S. and Hughes, M. K. (1999) Northern Hemisphere temperatures during the past millennium: inferences, uncertainties, and limitations. *Geophys. Res. Lett.*, **26**, 759–762.

Marshall, J. and Plumb, R. A. (2008) *Atmosphere Ocean, and Climate Dynamics: An Introductory Text.* International Geophysics Series, v. 93, Elsevier Academic Press, Burlington, MA. 319 pp.

Martinson, D. G., Pisias, N. G., Hays, J. D. *et al.* (1987) Age dating and the orbital theory of the ice ages; development of a high-resolution 0 to 300 000-year chronostratigraphy. *Quat. Res.*, **27**, 1–29.

Maslin, M. A., Seidov, D. and Lowe, J. (2001) Synthesis and nature and causes of rapid climate transitions during the Quaternary (a review). In *The Oceans and Rapid Climate Change*, ed. D. Seidov, B. J. Haupt, E. J. Barron and M. Maslin, *Amer. Geophys. U. Geophys. Monogr.*, **126**, 9–51.

Mattinson, J. M. (1978) Age, origin, and thermal histories of some plutonic rocks from Salinian Block of California. *Contrib. Mineral. Petrol.*, **67**, 233–245.

Mayewski, P. A., Meeker, L. D., Whitlow, S. *et al.* (1994) Changes in atmospheric circulation and ocean ice cover over the North Atlantic during the last 41 000 years. *Science*, **263**, 1747–1751.

McCreary, J. P., Kundu, P. K. and Molinari, R. L. (1993) A numerical investigation of dynamics, thermodynamics, and mixed layer processes in the Indian Ocean. *Prog. Ocean.*, **31**, 181–244.

McDonald, W. F. (1938) Atlas of climate charts of the oceans, Charts 59–62, Washington, DC: US Department of Agriculture, Weather Bureau, p. 247.

McIntyre, A. and Molfino, B. (1996) Forcing Atlantic equatorial and subpolar millennial cycles by precession. *Science*, **274**, 1867–1870.

Meehl, G. A. (1994) Coupled ocean–atmosphere–land processes and south Asian monsoon variability. *Science*, **265**, 263–267.

Meehl, G. A. (1997) The south Asian monsoon and the tropospheric biennial oscillation. *J. Clim.*, **10**, 1921–1943.

Mercier, J. L., Armijo, R., Tapponnier, P., Carey, G. E. and Han, T. L. (1987) Change from late Tertiary compression to Quaternary extension in southern Tibet during the India–Asia collision. *Tectonics*, **6**, 275–304.

Métivier, F., Gaudemer, Y., Tapponnier, P. and Klein, M. (1999) Mass accumulation rates in Asia during the Cenozoic. *Geophys. J. Int.*, **137**, 280–318.

Miller, K. G., Wright, J. M. and Fairbanks, R. G. (1991) Unlocking the Ice House: Oligocene–Miocene oxygen isotopes, eustasy, and margin erosion. *J. Geophys. Res.*, **96**, 6829–6848.

Milliman, J. D. and Syvitski, J. P. M. (1992) Geomorphic/tectonic control of sediment discharge to the ocean; the importance of small mountainous rivers. *J. Geology*, **100**, 525–544.

Mischke, S., Herzschuh, U., Zhang, C., Bloemendal, J. and Riedel, F. (2005) Late Quaternary lake record from the Qilian Mountains (NW China): lake level and salinity changes inferred from sediment properties and ostracod assemblages. *Global Planet. Change*, **46**, 337–359.

Moberg, A., Sonechkin, D. M., Holmgren, K., Datsenko, N. M. and Karlen, W. (2005) Highly variable Northern Hemisphere temperatures reconstructed from low and high-resolution proxy data. *Nature*, **433**, 613–617.

Molnar, P. (2004) Late Cenozoic increase in accumulation rates of terrestrial sediment: how might climate change have affected erosion rates? *Ann. Rev. Earth Planet. Sci.*, **32**, 67–89.

Molnar, P. and Emanuel, K. A. (1999) Temperature profiles in radiative-convective equilibrium above surfaces at different heights. *J. Geophys. Res.*, **104**, 24 265–24 484.

Molnar, P. and England, P. C. (1990) Late Cenozoic uplift of mountain ranges and global climate change; chicken or egg? *Nature*, **346**, 29–34.

Molnar, P., England, P. and Martinod, J. (1993) Mantle dynamics, the uplift of the Tibetan Plateau, and the Indian monsoon. *Rev. Geophys.*, **31**, 357–396.

Morgan, M. E., Kingston, J. D. and Marino, B. D. (1994) Carbon isotopic evidence for the emergence of C4 plants in the Neogene from Pakistan and Kenya. *Nature*, **367**, 162–165.

Mountain, G. S. and Prell, W. L. (1990) A multiphase plate tectonic history of the south-east continental margin of Oman. In *The Geology and Tectonics of the Oman Region*, ed. A. H. F. Robertson, M. P. Searle and A. C. Ries. Geol. Soc., Lond., Spec. Publ., pp. 725–743.

Murphy, M. A., Yin, A., Harrison, T. M. *et al.* (1997) Did the Indo–Asian collision alone create the Tibetan Plateau? *Geology*, **25**, 719–722.

Murthy, S. R. N. (1980) The Vedic River Saraswati, a myth or fact; a geological approach. *Indian J. Hist. Sci.*, **15** (2), 189–192.

Nair, R. R., Ittekkot, V., Manganini, S. J. *et al.* (1989) Increased particle flux to the deep ocean related to monsoons. *Nature*, **338**, 749–751.

Najman, Y. (2006) The detrital record of orogenesis: a review of approaches and techniques used in the Himalayan sedimentary basins. *Earth-Sci. Rev.*, **74**, 1–72.

Najman, Y., Pringle, M., Godin, L. and Oliver, G. (2001) Dating of the oldest continental sediments from the Himalayan foreland basin. *Nature*, **410**, 194–197.

Naqvi, W. A. and Fairbanks, R. G. (1996) A 27 000 year record of Red Sea outflow; implication for timing of post-glacial monsoon intensification. *Geophys. Res. Lett.*, **23**, 1501–1504.

Nathan, S. A. and Leckie, R. M. (2004) Gateway closures and ocean circulation during the late Miocene (approximately 13–5 Ma). *Abstr. Prog., Geol. Soc. Amer.*, **36**, 197.

Nathan, S. A., Leckie, R. M., Olson, B. and Deconto, R. M. (2003) The Western Pacific Warm Pool; a probe of global sea level change and Indonesian Seaway closure during the middle to late Miocene. *Ann. Meet. Abstr., Amer. Assoc. Petrol. Geol.*, **12**, 126.

Neelin, J. D. (2007) Moist dynamics of tropical convection zones in monsoons, teleconnections and global warming. In *The Global Circulation of the Atmosphere*, ed. T. Schneider and A. Sobel, Princeton, NJ: Princeton University Press, pp. 267–301.

Neff, U., Burns, S. J., Mangini, A. *et al.* (2001) Strong coherence between solar variability and the monsoon in Oman between 9 and 6 kyr ago. *Nature*, **411**, 290–293.

Nelson, K. D., Zhao, W., Brown, L. D. *et al.* (1996) Partially molten middle crust beneath southern Tibet: synthesis of Project INDEPTH results. *Science*, **274**, 1684–1688.

Nesbitt, G. M. and Young, H. W. (1982) Early Proterozoic climates and plate motions inferred from major element chemistry of lutites. *Nature*, **299**, 715–717.

Nicholls, N. (1983) Air–sea interaction and the quasi-biennial oscillation. *Mon. Wea. Rev.*, **106**, 1505–1508.

Nitsuma, N., Oba, T. and Okada, M. (1991) Oxygen and carbon isotope stratigraphy at site 723, Oman Margin. *Proc. Ocean Drilling Prog., Sci. Res.*, **117**, College Station, TX: Ocean Drilling Program, 321–341.

Oldham, R. D. (1893) The Saraswati and the Lost River of the Indian Desert. *J. R. Asiatic Soc.*, **1893**, 49–76.

Oppo, D. W. and Lehman, S. J. (1995) Suborbital timescale variability of North Atlantic Deep Water during the past 200 000 years. *Paleoceanography*, **10**, 901–910.

Oppo, D. W., Linsley, B. K., Rosenthal, Y., Dannenmann, S. and Beaufort, L. (2003) Orbital and suborbital climate variability in the Sulu Sea, western tropical Pacific. *Geochem., Geophys., Geosyst.*, **1003**, doi: 10.1029/2001GC000260.

Overpeck, J. and Cole, J. E. (2007) Climate change: lessons from a distant monsoon. *Nature*, **445**, 270–271.

Overpeck, J., Anderson, D., Trumbore, S. and Prell, W. (1996a) The south-west Indian Monsoon over the last 18 000 years. *Clim. Dynam.*, **12**, 213–225.

Overpeck, J., Rind, D., Lacis, A. and Healy, R. (1996b) Possible role of dust-induced regional warming in abrupt climate change during the last glacial period. *Nature*, **384**, 447–449.

Pairault, A. A., Hall, R. and Elders, C. F. (2003) Structural styles and tectonic evolution of the Seram Trough, Indonesia. *Mar. Petrol. Geol.*, **20**, 1141–1160.

Palmer, M. R. and Edmond, J. M. (1989) The strontium isotope budget of the modern ocean. *Earth Planet. Sci. Lett.*, **92**(1), 11–26.

Pan, Y. and Kidd, W. S. F. (1992) Nyainqentanglha shear zone; a late Miocene extensional detachment in the southern Tibetan Plateau. *Geology*, **20**, 775–778.

Pang, K. D. (1987) Extraordinary floods in early Chinese history and their absolute dates. *J. Hydrol.*, **96**, 139–155.

Pearson, P. N. and Palmer, M. R. (2000) Atmospheric carbon dioxide concentrations over the past 60 million years. *Nature*, **406**, 695–699.

Pechenkina, E. A, Benfer, R. A. and Wang, Z. (2001) Diet and health changes at the end of the Chinese neolithic: the Yangshao/Longshan transition in Shaanxi province. *Amer. J. Phys. Anthrop.*, **117**, 15–36.

Pelejero, C. and Grimalt, J. O. (1997) The correlation between the UK37 index and sea-surface temperature in the warm boundary: the South China Sea. *Geochim. Cosmochim. Acta*, **61**, 4789–4797.

Peltier, W. R. (1994) Ice age paleotopography. *Science*, **265**, 195–201.

Peregrine, P. (1991) Some political aspects of craft specialization. *World Archaeo.*, **23**, 1–11.

Peterson, L. C., Murray, D. W., Ehrmann, W. U. and Hempel, P. (1992) Cenozoic carbonate accumulation and compensation depth changes in the Indian Ocean. In: *Synthesis of Results From Scentific Drilling in the Indian Ocean*, ed. R. A. Duncan, D. K. Rea, R. B. Kidd, U. von Rad and J. K. Weissel, American Geophysical Union Monog. **70**, 311–333.

Petit, J. R., Jouzel, J., Raynaud, D. *et al.* (1999) Climate and atmospheric history of the past 420 000 years from the Vostok ice core, Antarctica. *Nature*, **399**, 429–436.

Petit, J. R., Jouzel, J., Raynaud, D. *et al.* (2001) *Vostok Ice Core Data for 420 000 Years*, IGBP PAGES/World Data Center for Paleoclimatology Data Contribution Series #2001-076. Boulder CO, USA: NOAA/NGDC Paleoclimatology Program.

Pettke, T., Halliday, A. N., Hall, C. M. and Rea, D. K. (2000) Dust production and deposition in Asia and the North Pacific Ocean over the past 12 Myr. *Earth Planet. Sci. Lett.*, **178**, 397–413.

Porter, S. C. and An, Z. S. (1995) Correlation between climate events in the North Atlantic and China during the last glaciation. *Nature*, **375**, 305–308.

Possehl, G. (1990) Revolution in the urban revolution: the emergence of Indus urbanization. *Ann. Rev. Anthrop.*, **19**, 261–282.

Possehl, G. (1993) *Harappan Civilization: A Recent Perspective*. Delhi: Oxford University Press, p. 595.

Possehl, G. (1997) Climate and the eclipse of the ancient cities of the Indus. In *Third Millennium BCE Climate Change and Old World Collapse*, ed. H. N. Dalfes, G. Kukla and H. Weiss, NATO ASI Ser. 1, **49**, New York: Springer, pp. 193–244.

Potts, D. T. (1999) *The Archaeology of Elam: Formation and Transformation of an Ancient Iranian State*. Cambridge: Cambridge University Press, p. 488.

Powell, C. M. (1986) Curvature of the Himalayan Arc related to Miocene normal faults in southern Tibet. *Geology*, **14**, 358–359.

Prabhu, C. N., Shankar, R., Anupama, K. *et al.* (2004) A 200-ka pollen and oxygen-isotopic record from two sediment cores from the eastern Arabian Sea. *Palaeogeog., Palaeoclim., Palaeoeco.*, **214**, 309–321.

Prawdin, M. (2006) *The Mongol Empire: Its Rise and Legacy*. New York: Transaction Publishers, p. 581.

Prell, W. L. and Curry, W. B. (1981) Faunal and isotopic indices of monsoonal upwelling: western Arabian Sea. *Ocean. Acta*, **4**, 91–98.

Prell, W. L. and Kutzbach, J. E. (1987) Monsoon variability over the past 150 000 years. *J. Geophys. Res. Atmos. Sci.*, **92** (D7), 8411–8425.

Prell, W. L. and Kutzbach, J. E. (1992) Sensitivity of the Indian monsoon to forcing parameters and implications for its evolution. *Nature*, **360**, 647–652.

Prell, W. L, Murray, D. W., Clemens, S. C. and Anderson, D. M. (1992) Evolution and variability of the Indian Ocean Summer Monsoon: evidence from the western Arabian Sea drilling program. In *Synthesis of Results from Scientific Drilling in the Indian Ocean,* ed. R. A. Duncan, D. K. Rea, R. B. Kidd, U. von Rad and J. K. Weissel. *Amer. Geophys. U. Monogr.*, 70, 447–469.

Prins, M. A. and Postma, G. (2000) Effects of climate, sea level, and tectonics unraveled for last deglaciation turbidite records of the Arabian Sea. *Geology*, **28**, 375–378.

Privé, N. C. and Plumb, R. A. (2007a) Monsoon dynamics with interactive forcing. Part I: axisymmetric studies. *J. Atmos. Sci.*, **64** (5), 1417–1430.

Privé, N. C. and Plumb, R. A. (2007b) Monsoon dynamics with interactive forcing. Part II: impact of eddies and asymmetric geometries. *J. Atmos. Sci.*, **64** (5), 1431–1442.

Prospero, J. M., Uematsi, M. and Savoie, D. L. (1989) Mineral aerosol transport to the Pacific Ocean. In *Chemical Oceanography*, vol. 10, ed. J. P. Riley, R. Chester and R. A. Duce, San Diego: Academic Press, pp. 187–218.

Purdy, J. and Jäger, E. (1976) K-Ar ages on rock-forming minerals from the Central Alps. *Mem. 1st Geol. Min. Congr., Univ. Padova*, **30**, p. 32.

Puri, V. M. K. (2001) Origin and course of Vedic Saraswati River in Himalaya; its secular desiccation episodes as deciphered from palaeo-glaciation and geomorphological signatures. In *Proceedings; Symposium on Snow, Ice and Glaciers; a Himalayan Perspective*, ed. S. K. Acharyya. *Geol. Surv. India, Spec. Publ. Ser.*, **53**, 175–191.

Pye, K. and Zhou, L. (1989) Late Pleistocene and Holocene aeolian dust deposition in north China and the North-west Pacific Ocean. *Palaeogeog., Palaeoclim., Palaeoeco.*, **73**, 11–23.

Quade, J. (1993) Major shifts in the $^{87}Sr/^{86}Sr$ ratios of large paleorivers draining the Himalayas of central Nepal over the past 10 Ma. *Geol. Soc. America, Abstr. Prog.*, **25** (6), 175.

Quade, J., Cerling, T. E. and Browman, J. R. (1989) Dramatic ecologic shift in the late Miocene of northern Pakistan, and its significance to the development of the Asian monsoon. *Nature*, **342**, 163–166.

Ramage, C. (1971) *Monsoon Meteorology*. New York: Academic Press, *International Geophysics Series*, vol. **15**, p. 296.

Ramstein, G., Fluteau, F., Besse, J. and Joussaume, S. (1997) Effect of orogeny, plate motion and land–sea distribution on Eurasian climate change over the past 30 million years. *Nature*, **386**, 788–795.

Rao, Y. P. (1976) *South-west Monsoon*. New Delhi: India Meteorological Department, p. 367.

Raval, A. and Ramanathan, V. (1989) Observational determination of the greenhouse effect. *Nature*, **342**, 758–762.

Raymo, M. E. and Ruddiman, W. F. (1992) Tectonic forcing of the late Cenozoic climate. *Nature*, **359**, 117–122.

Raymo, M. E., Ruddiman, W. F. and Froelich, P. (1988) Influence of late Cenozoic mountain building on ocean geochemical cycles. *Geology*, **16**, 649–653.

Rea, D. K. (1992) Delivery of Himalayan sediment to northern Indian Ocean and its relation to global climate, sea level, uplift and seawater strontium. In *Synthesis of Results from Scientific Drilling in the Indian Ocean*, ed. R. A. Duncan, D. K. Rea, R. B. Kidd, U. von Rad and J. K. Weissel. *Amer. Geophys. U. monogr.*, **70**, 387–402.

Rea, D. K. (1994) The paleoclimatic record provided by Eolian deposition in the deep sea; the geologic history of wind. *Rev. Geophys.*, **32**, 159–195.

Rea, D. K., Basov, I. A., Janecek, T. R. *et al.* (1993) *Proc. Ocean Drill. Prog. Init. Rpts*, **145**, College Station, TX: Ocean Drilling Program, p. 1040.

Rea, D. K., Snoeckx, H and Joseph, L. H. (1998) Late Cenozoic Eolian deposition in the North Pacific: Asian drying, Tibetan uplift, and cooling of the northern hemisphere. *Paleoceanography*, **13**, 215–224.

Reade, J. (2001) Assyrian king-lists, the royal tombs of Ur, and Indus origins. *J. Near East. Stud.*, **60**, 1–29.

Redfield, C. C., Ketchum, B. H. and Richards, F. A. (1963) The influence of organisms on the composition of sea-water. In *The Sea*, ed. M. N. Hill, New York: Wiley-Interscience, pp. 26–77.

Reichart, G. J., den Dulk, M., Visser, H. J., van der Weijden, C. H. and Zachariasse, W. J. (1997) A 225 kyr record of dust supply, paleoproductivity and the oxygen minimum zone from the Murray Ridge (northern Arabian Sea). *Palaeogeog. Palaeoclim. Palaeoeco.*, **134**, 149–169.

Reiners, P. W., Ehlers, T. A., Mitchell, S. G. and Montgomery, D. R. (2003) Coupled spatial variations in precipitation and long-term erosion rates across the Washington Cascades. *Nature*, **426**, 645–647.

Ren, S. N. (2000) The origin and development of settlement and society during the Neolithic Age in China. *Archaeology*, **7**, 48–59 (in Chinese).

Richards, J. F. (1990) The seventeenth-century crisis in South Asia. *Mod. Asian Stud.*, **24**, 625–638.

Richards, J. F. (1996) *The Mughal Empire*. Cambridge: Cambridge University Press, p. 337.

Richerson, P. J., Boyd, R. and Bettinger, R. L. (2001) Was agriculture impossible during the Pleistocene but mandatory during the Holocene? A climate change hypothesis. *Amer. Antiq.*, **66** (3), 387–412.

Rodgers, D. W. and Gunatilaka, A. (2002) Bajada formation by monsoonal erosion of a subaerial forebulge, Sultanate of Oman. *Sed. Geol.*, **154**, 127–146.

Rodwell, M. J. and Hoskins, B. J. (1995) A model of the Asian summer monsoon. Part II: cross-equatorial flow and PV behavior. *J. Atmos. Sci.*, **52**, 1341–1356.

Rodwell, M. J. and Hoskins, B. J. (2001) Subtropical anticyclones and summer monsoons. *J. Clim.*, **14**, 3192–3211.

Rogalla, U. and Andruleit, H. (2005) Precessional forcing of coccolithophore assemblages in the northern Arabian Sea; implications for monsoonal dynamics during the last 200 000 years. *Mar. Geol.*, **217**, 31–48.

Rögl, F. and Steininger, F. F. (1984) Neogene Paratethys, Mediterranean and Indo-pacific seaways. In *Fossils and Climate*, ed. P. J. Brenchley, New York: John Wiley and Sons., pp. 171–200.

Rossignol-Strick, M., Paterne, M., Bassinot, F. C., Emeis, K. C. and de Lange, G. J. (1998) An unusual mid-Pleistocene monsoon period over Africa and Asia. *Nature*, **392**, 269–272.

Roth, J. M., Droxler, A. W. and Kameo, K. (2000) The Caribbean carbonate crash at the middle to late Miocene transition: linkage to the establishment of the modern global ocean conveyor. *Proc. Ocean Drill. Prog., Sci. Res.*, **165**, College Station, TX: Ocean Drilling Program, pp. 249–273.

Rowley, D. B. (1996) Age of initiation of collision between India and Asia; a review of stratigraphic data, *Earth Planet. Sci. Lett.*, **145**, 1–13.

Rowley, D. B. and Currie, B. S. (2006) Palaeo-altimetry of the late Eocene to Miocene Lunpola basin, central Tibet. *Nature*, **439**, 677–681.

Rowley, D. B., Pierrehumbert, R. T. and Currie, B. S. (2001) A new approach to stable isotope-based paleoaltimetry: implications for paleoaltimetry and paleohypsometry of the High Himalaya since the Late Miocene. *Earth Planet. Sci. Lett.*, **188**, 253–268.

Ruddiman, W. F. and Kutzbach, J. E. (1989) Forcing of late Cenozoic Northern Hemisphere climate by plateau uplift in southern Asia and the American West. *J. Geophys. Res.*, **94** (15), 18 409–18 427.

Rutherford, E., Burke, K. and Lytwyn, J. (2001) Tectonic history of Sumba Island, Indonesia, since the Late Cretaceous and its rapid escape into the forearc in the Miocene. *J. Asian Earth Sci.*, **19**, 453–479.

Saito, Y. (1998) Sea levels of the last glacial in the East China Sea continental shelf. *Quat. Res.*, **37**, 235–242 (in Japanese with English abstract).

Saji, N. H., Goswami, B. N., Vinayachandran, P. N. and Yamagata, T. (1999) A dipole mode in the tropical Indian Ocean. *Nature*, **401**, 360–363.

Sanders, F. (1984) Quasi-geostrophic diagnosis of the Monsoon Depression of 5–8 July, 1979. *J. Atmos. Sci.* **41**, 538–552.

Schmitz, W. J. and McCartney, M. S. (1993) On the North Atlantic circulation. *Rev Geophys.*, **31**, 29–50.

Schoenbohm, L. M., Burchfiel, B. C., Chen, L. and Yin, J. (2006) Miocene to present activity along the Red River fault, China, in the context of continental extrusion, upper-crustal rotation, and lower-crustal flow. *Geol. Soc. Amer. Bull.*, **118**, 672–688.

Schwenk, T., Spiess, V., Hübscher, C. and Breitzke, M. (2003) Frequent channel avulsions within the active channel–levee system of the middle Bengal Fan – an exceptional channel–levee development derived from Parasound and Hydrosweep data. *Deep-Sea Res. II*, **50**, 1023–1045.

Searle, M. P. and Godin, L. (2003) The South Tibetan detachment and the Manaslu Leucogranite; a structural reinterpretation and restoration of the Annapurna–Manaslu Himalaya, Nepal. *J. Geology*, **111**, 505–523.

Shackleton, N. J. and Opdyke, N. D. (1977) Oxygen isotope and palaeomagnetic evidence for early Northern Hemisphere glaciation. *Nature*, **270**, 216–219.

Shackleton, N. J., Berger, A. and Peltier, W. A. (1990) An alternative astronomical calibration of the lower Pleistocene timescale based on ODP Site 677. *Trans. R. Soc. Edin. Earth Sci.*, **81**, 251–261.

Shackleton, N. J., Hall, M. A. and Pate, D. (1995) Pliocene stable isotope stratigraphy of ODP Site 846. In: *Proc. Ocean Drill Prog., Sci. Rpt*, ed. N. G. Pisias *et al.*, 337–355.

Shaffer, J. G. (1992) The Indus Valley, Baluchistan and Helmand Traditions: Neolithic through Bronze Age. In *Chronologies in Old World Archaeology*, ed. R. W. Ehrich, Chicago: University of Chicago Press, pp. 425–446.

Sharma, S., Joachimski, M., Sharma, M. *et al.* (2004) Late glacial and Holocene environmental changes in Ganga plain, northern India. *Quat. Sci. Rev.*, **23**, 145–159.

Sheppard, P. R., Tarasov, P. E., Graumlich, L. J. *et al.* (2004) Annual precipitation since 515 BCE reconstructed from living and fossil juniper growth of north-eastern Qinghai Province, China. *Clim. Dynam.*, **23**, 869–881.

Shui T. (2001) *Papers on the Bronze Age Archaeology of North-west.* Beijing: China Science Press, p. 25 (in Chinese).

Shukla, J. and Paolina, D. A. (1983) The Southern Oscillation and long range forecasting of the summer monsoon rainfall over India. *Mon. Wea. Rev.*, **111**, 1830–1837.

Siddall, M., Rohling, E. J., Almogi-Labin, A. *et al.* (2003) Sea-level fluctuations during the last glacial cycle. *Nature*, **423**, 583–588.

Sikka, D. R. and Gadgil, S. (1980) On the Maximum Cloud Zone and the ITCZ over Indian longitudes during the South-west Monsoon. *Mon. Wea. Rev.*, **108**, 1840–1853.

Sinha, A., Cannariato, K. G., Stott, L. D. *et al.* (2005) Variability of South-west Indian summer monsoon precipitation during the Bølling-Allerød. *Geology*, **33**, 813–816.

Sirocko, F. (1995) Abrupt change in monsoonal climate: evidence from the geochemical composition of Arabian Sea sediments. Habilitation Thesis, University of Kiel, p. 216.

Sirocko, F., Sarnthein, M., Erlenkeuser, H. *et al.* (1993) Century-scale events in monsoonal climate over the last 24 000 years. *Nature*, **364**, 322–324.

Sirocko, F., Garbe-Schönberg, D., McIntyre, A. and Molfino, B. (1996) Teleconnections between the subtropical monsoons and high-latitude climates during the last deglaciation. *Science*, **272**, 526–529.

Sirocko, F., Garbe-Schoenberg, D. and Devey, C. (2000) Processes controlling trace element geochemistry of Arabian Sea sediments during the last 25 000 years. *Global Planet. Change*, **26** (1–3), 217–303.

Sontakke, N. A., Pant, G. B. and Singh, N. (1993) Construction of all-India summer monsoon rainfall series for the period 1844–1991. *J. Climat.*, **6**, 1807–1811.

Spicer, R. A., Harris, N. B. W., Widdowson, M. *et al.* (2003) Constant elevation of southern Tibet over the past 15 million years. *Nature*, **421**, 622–624.

Staubwasser, M., Sirocko, F., Grootes, P. M. and Segl, M. (2003) Climate change at the 4.2 ka BP termination of the Indus valley civilization and Holocene south Asian monsoon variability. *Geophys. Res. Lett.*, **30** (8), 1425, doi: 10.1029/2002GL016822.

Stevens, T., Armitage, S. J., Lu, H. and Thomas, D. S. G. (2006) Sedimentation and diagenesis of Chinese loess: implications for the preservation of continuous, high-resolution climate records. *Geology*, **34**, 849–852.

Stott, L., Poulsen, C., Lund, S. and Thunell, R. (2002) Super ENSO and global climate oscillations at millennial timescales. *Science*, **297**, 222–226.

Stuiver, M. and Braziunas, T. F. (1993) Sun, ocean, climate and atmospheric $^{14}CO_2$; an evaluation of causal and spectral relationships. *The Holocene*, **3**, 289–305.

Stuiver, M. and Grootes, P. M. (2000) GISP2 oxygen isotope ratios. *Quat. Res.*, **53**, 277–284.

Sun, D. (2004) Monsoon and westerly circulation changes recorded in the late Cenozoic aeolian sequences of northern China. *Global Planet. Change*, **41**, 63–80.

Sun, D. H., Liu, T. S., Chen, M. Y. and An, Z. S. (1997) Magnetostratigraphy and climate implications of the Red–Clay sequences in the Loess Plateau in China. *Sci. in China*, **27**, 265–270.

Sun, D. H., Shaw, J., An, Z. S., Chen, M. Y. and Yue, L. P. (1998) Magnetostratigraphy and paleoclimatic interpretation of continuous 7.2 Ma late Cenozoic Eolian sediments from the Chinese Loess Plateau. *Geophys. Res. Lett.*, **25**, 85–88.

Sun, X. and Li, X. (1999) Pollen records of the last 37 ka in deep-sea core 17940 from the northern slope of the South China Sea. *Mar. Geol.*, **156**, 227–244.

Sun, X. and Wang, P. (2005) How old is the Asian monsoon system? Palaeobotanical records from China. *Palaeogeog., Palaeoclim., Palaeoeco.*, **222**, 181–222.

Sun, Y. and An, Z. (2004) An improved comparison of Chinese loess with deep-sea $\delta^{18}O$ record over the interval 1.6–2.6 Ma. *Geophys. Res. Lett.*, **31**, 13.

Suzuki, H. (1978) *Ideas of the Forest and Ideas of the Desert*. Tokyo: NHK Books, p. 222.

Swain, A. M., Kutzbach, J. E. and Hastenrath, S. (1983) Estimates of Holocene precipitation for Rajasthan, India, based on pollen and lake-level data. *Quat. Res.*, **19**, 1–17.

Tada, R., Irino, T. and Koizumi, I. (1999) Land–ocean linkages over orbital and millennial timescales recorded in late Quaternary sediments of the Japan Sea. *Paleoceanography*, **14**, 236–247.

Tamburini, F., Adatte, T., Foellmi, K., Bernasconi, S. M. and Steinmann, P. (2003) Investigating the history of East Asian monsoon and climate during the last glacial–interglacial period (0–140,000 years); mineralogy and geochemistry of ODP Sites 1143 and 1144, South China Sea. *Mar. Geol.*, **201**, 147–168.

Tapponnier, P., Xu Z., Roger, F. *et al.* (2001) Oblique stepwise rise and growth of the Tibet Plateau. *Science*, **294**, 1671–1677.

Thamban, M., Rao, V. P., Schneider, R. R. and Grootes, P. M. (2001) Glacial to Holocene fluctuations in hydrography and productivity along the south-western continental margin of India. *Palaeogeogr., Palaeoclimatol., Palaeoecol.*, **165**, 113–127.

Thiede, R. C., Bookhagen, B., Arrowsmith, J. R., Sobela, E. R. and Strecker, M. R. (2004) Climatic control on rapid exhumation along the Southern Himalayan Front. *Earth Planet. Sci. Lett.*, **222**, 791–806.

Thiry, M. (2000) Palaeoclimatic interpretation of clay minerals in marine deposits: an outlook from the continental origin. *Earth Sci. Rev.*, **49**, 201–221.

Thompson, L. G., Mosley-Thompson, E., Davis, M. E. *et al.* (1989) 100,000 year climate record from Qinghai–Tibetan Plateau ice cores. *Science*, **246**, 474–477.

Thompson, L. G., Mosley-Thompson, E., Davis, M. *et al.* (1993) "Recent warming"; ice core evidence from tropical ice cores with emphasis on Central Asia. *Global Planet. Change*, **7**, 145–156.

Thompson, L. G., Yao, Y., Davis, M. E. *et al.* (1997) Tropical climate instability: the last glacial cycle from a Qinghai–Tibetan ice core. *Science*, **276**, 1821–1825.

Thompson, L. G., Yao, T., Mosley-Thompson, E. *et al* (2000) A high-resolution millennial record of the South Asian Monsoon from Himalayan ice cores. *Science*, **289**, 1916–1919.

Tomczak, M. and Godfrey. J. S. (1994) *Regional Oceanography: An Introduction*. Oxford: Pergamon Press, p. 422.

Torrence, C. and Webster, P. J. (1999) Interdecadal changes in the ENSO–Monsoon System. *J. Clim.*, **12**, 2679–2690.

Tosi, M. (1975) The dialectics of state formation in Mesopotamia, Iran, and central Asia. *Dialect. Anthrop.*, **1**, 173–180.

Treloar, P. J., Rex, D. C., Guise, P. G. *et al.* (1989) K-Ar and Ar-Ar geochronology of the Himalayan collision in NW Pakistan: constraints on the timing of suturing, deformation, metamorphism and uplift. *Tectonics*, **8**, 881–909.

Treydte, K. S., Schleser, G. H., Helle, G. *et al.* (2006) The twentieth century was the wettest period in northern Pakistan over the past millennium. *Nature*, **440**, 1179–1182.

Turner, S., Hawkesworth, C., Liu, J. *et al.* (1993) Timing of Tibetan uplift constrained by analysis of volcanic rocks. *Nature*, **364**, 50–54.

Underhill, A. P. (1991) Pottery production in chiefdoms: the Longshan period in northern China. *World Archaeo.*, **23**, 12–27.

Underhill, P. A., Passarino, G., Lin, A. A. *et al* (2001) The phylogeography of Y chromosome binary haplotypes and the origins of modern human populations. *Ann. Human Genet.*, **65**, 43–62.

Vail, P. R., Mitchum, R. M. and Thompson, S. (1977) Seismic stratigraphy and global changes of sea level; Part 4; global cycles of relative changes of sea level. In *Seismic Stratigraphy; Applications to Hydrocarbon Exploration*, ed. C. E. Payton. *Mem., Amer. Assoc. Petrol. Geol.*, **26**, 83–97.

Van Campo, E. (1991) Pollen transport into Arabian Sea sediments. *Proc. Ocean Drill. Prog., Sci. Res.*, **117**, College Station, TX: Ocean Drilling Program, 277–281.

Van Campo, E., Duplessy, J. C. and Rossignol-Strick, M. (1982) Climatic conditions deduced from 150 000 yr oxygen isotope–pollen record from the Arabian sea. *Nature*, **296**, 56–59.

Vannay, J. C., Sharp, Z. D. and Grasemann, B. (1999) Himalayan inverted metamorphism constrained by oxygen isotope thermometry. *Contrib. Mineral. Petrol.*, **137**, 90–101.

Von Rad, U., Schulz, H., Riech, V. *et al.* (1999) Multiple monsoon-controlled breakdown of oxygen-minimum conditions during the past 30 000 years documented in laminated sediments off Pakistan. *Palaeogeog., Palaeoclim., Palaeoeco.*, **152**, 129–161.

Walker, C. B., Searle, M. P. and Waters, D. J. (2001) An integrated tectonothermal model for the evolution of the High Himalaya in western Zanskar with constraints from thermobarometry and metamorphic modeling. *Tectonics*, **20**, 810–833.

Wang, B., Clemens, S. and Liu, P. (2003a) Contrasting the Indian and East Asian monsoons: implications on geologic timescale. *Mar. Geol.*, **201**, 5–21.

Wang, J., Wang, Y., Liu, Z., Li, J. and Xi, P. (1999c) Cenozoic environmental evolution of the Qaidam Basin and its implications for the uplift of the Tibetan Plateau and the drying of Central Asia. *Palaeogeog., Palaeoclim., Palaeoeco.*, **152**, 37–47.

Wang, L., Sarnthein, M., Erlenkeuser, H. *et al.* (1999a) East Asian monsoon climate during the late Pleistocene: high-resolution sediment records from the South China Sea. *Mar. Geol.*, **156**, 245–284.

Wang, L., Sarnthein, M., Erlenkeuser, H. *et al.* (1999b) Holocene variations in Asian monsoon moisture; a bidecadal sediment record from the South China Sea. *Geophys. Res. Lett.*, **26**, 2889–2892.

Wang, P. (2004) Cenozoic deformation and the history of sea–land interactions in Asia. In *Continent–Ocean Interactions Within East Asian Marginal Seas*, ed. P. D. Clift, W. Kuhnt, P. Wang and D. E. Hayes, Washington D. C.: American Geophysical Union, *Geophys. Monogr. Ser.* **149**, pp. 1–22.

Wang, P., Zhao, Q., Jian, Z. *et al.* (2003b). Thirty million year deep-sea records in the South China Sea. *Chinese Sci. Bull.*, **48**, 2524–2535.

Wang, P., Tian J., Cheng X., Liu C. and Xu J. (2003c) Carbon reservoir changes preceded major ice-sheet expansion at the mid-Brunhes event. *Geology*, **31**, 239–242.

Wang, R. and Abelmann, A. (2002) Radiolarian responses to paleoceanographic events of the southern South China Sea during the Pleistocene. *Mar. Micropaleo.*, **46**, 25–44.

Wang, R. L., Scarpitta, S. C., Zhang, S. C. and Zheng, M. P. (2002) Later Pleistocene/Holocene climate conditions of Qinghai–Xizhang Plateau (Tibet) based on carbon and oxygen stable isotopes of Zabuye Lake sediments. *Earth Planet. Sci. Lett.*, **203**, 461–477.

Wang, Y. J., Cheng, H., Edwards, R. L. *et al.* (2001a) A high-resolution absolute-dated late Pleistocene Monsoon record from Hulu Cave, China. *Science*, **294**, 2345–2348.

Wang, Y., Cheng, H., Edwards, R. L. *et al.* (2005) The Holocene Asian monsoon; links to solar changes and North Atlantic climate. *Science*, **308**, 854–857.

Wang, Y., Deng, T. and Biasatti, D. (2006) Ancient diets indicate significant uplift of southern Tibet after ca. 7 Ma. *Geology*, **34**, 309–312.

Weber, M. E., Wiedicke, M. H., Kudrass, H. R., Huebscher, C. and Erlenkeuser, H. (1997) Active growth of the Bengal Fan during sea-level rise and highstand. *Geology*, **25**, 315–318.

Webster, P. J. (1987) The elementary monsoon. In *Monsoons*, ed. J. S. Fein and P. L. Stephens, New York: John Wiley, pp. 3–32.

Webster, P. J., Magana, V. O., Palmer, T. N. *et al.* (1998) Monsoons: processes, predictability, and the prospects for prediction, in the TOGA decade. *J. Geophys. Res.*, **103**, 14 451–14 510.

Webster, P. J., Clark, C., Cherikova, G. *et al.* (2002) The Monsoon as a self-regulating coupled ocean-atmosphere system. In *Meteorology at the Millennium,* ed. R. P. Pearce San Diego: Academic Press, *International Geophysical Series*, vol. **83**, 198–219.

Webster, P. J., Holland, G. J., Curry, J. A. and Chang, H.-R. (2005) Changes in tropical cyclone number, duration, and intensity in a warming environment. *Science*, **309** (5742), 1844–1846.

Weiss H, Courty M. A. and Wetterstrom W. (1993) The genesis and collapse of third millennium north Mesopotamian civilization. *Science*, **261**, 995–1004.

Whipple, K. X., Kirby, E. and Brocklehurst, S. H. (1999) Geomorphic limits to climate-induced increases in topographic relief. *Nature*, **401**, 39–43.

White, N. M., Pringle, M., Garzanti, E. *et al.* (2002) Constraints on the exhumation and erosion of the High Himalayan Slab, NW India, from foreland basin deposits. *Earth Planet. Sci. Lett.*, **195**, 29–44.

Williams, H., Turner, S., Kelley, S. and Harris, N. (2001) Age and compositin of dikes in southern Tibet: new constraints on the timing of east–west extension and its relations to postcollisional volcanism. *Geology*, **29**, 339–342.

Williams, H. M., Turner, S. P., Pearce, J. A., Kelley, S. P. and Harris, N. B. W. (2004) Nature of the source regions for post-collisional, potassic magmatism in Southern and Northern Tibet from geochemical variations and inverse trace element modeling. *J. Petrol.*, **45**, 555–607.

Wilson, R. J. S., Luckman, B. H. and Esper, J. A. (2005) 500-year dendroclimatic reconstruction of spring/summer precipitation from the lower Bavarian forest region, Germany. *Int. J. Climat.*, **25**, 611–630.

Wobus, C., Heimsath, A., Whipple, A. and Hodges, K. (2005) Active out-of-sequence thrust faulting in the central Nepalese Himalaya. *Nature*, **434**, 1008–1011.

Wolfe, J. A., Forest, C. E. and Molnar, P. (1998) Paleobotanical evidence of Eocene and Oligocene paleoaltitudes in midlatitude western North America. *Geol. Soc. Amer. Bull.*, **110**, 664–678.

Woodruff, F. and Savin, S. (1989) Miocene deepwater oceanography. *Paleoceanography*, **4**, 87–140.

Woodruff, F. and Savin, S. M. (1991) Mid-Miocene isotope stratigraphy in the deep-sea: high resolution correlations, paleoclimatic cycles, and sediment preservation. *Paleoceanography*, **6**, 755–806.

Woodruff, F., Savin, S. M. and Abel, L. (1990) Miocene benthic foraminifer oxygen and carbon isotopes, Site 709, Indian Ocean. *Proc. Ocean Drill. Prog., Sci. Res.*, 115, College Station, TX: Ocean Drilling Program, pp. 519–528.

Wu, G., Pan, B., Guan, Q. and Xia, D. (2005) Terminations and their correlation with insolation in the Northern Hemisphere; a record from a loess section in north-west China. *Palaeogeog., Palaeoclim., Palaeoeco.*, **216**, 267–277.

Xiao, J. L. and An, Z. S. (1999) Three large shifts in East Asian monsoon circulation indicated by loess–Paleosol sequences in China and late Cenozoic deposits in Japan. *Palaeogeog., Palaeoclim., Palaeoeco.*, **154**, 179–189.

Xiao, J. L., Inouchi, Y., Kumai, H. *et al.* (1997) Eolian quartz flux to Lake Biwa, central Japan over the past 145 000 years. *Quat. Res.*, **48**, 48–57.

Xiao, J. L., An, Z. S., Liu, T. S. *et al.* (1999) East Asian monsoon variation during the last 130 000 years; evidence from the Loess Plateau of central China and Lake Biwa of Japan. *Quat. Sci. Rev.*, **18**, 147–157.

Xie, P. and Arkin, P. A. (1997) Global precipitation: a 17-year monthly analysis based on gauge observations, satellite estimates and numerical model outputs. *Bull. Amer. Meteo. Soc.*, **78**, 2539–2558.

Xie, S.-P., Xu, H., Saji, N. H., Wang, Y. and Liu, W. T. (2006) Role of narrow mountains in large-scale organization of Asian monsoon convection. *J. Clim.*, **19**, 3420–3429.

Xu, R. (1979) Discovery of *Glossopteris* flora in southern Tibet and its significance to geology and paleogeography. In *A Report of the Scientific Expedition in the Mount Everest Region; 1975*, Beijing, China: Sci. Press, pp. 77–88.

Yan, F., Ye, Y. and Mai, X. (1983) The sporo-pollen assemblage in the Luo4 drilling of Lop Lake in Uygur Autonomous Region of Xinjiang and its significance. *Seismo. Geo.*, **5**, 75–80 (in Chinese).

Yancheva, G., Nowaczyk, N. R., Mingram, J. *et al.* (2007) Influence of the intertropical convergence zone on the East Asian monsoon. *Nature*, **445**, 74–77.

Yang, F., Ma, Z. Q., Xu, T. C. and Ye, S. T. (1992) A Tertiary paleomagnetic stratigraphic profile in Qaidam basin (in Chinese). *Acta Pet. Sin.*, **13** (2), 97–101.

Yang, J., Chen, J., An, Z. *et al.* (2000) Variations in $^{87}Sr/^{86}Sr$ ratios of calcites in Chinese loess; a proxy for chemical weathering associated with the East Asian summer monsoon. *Palaeogeog., Palaeoclim., Palaeoeco.*, **157**, 151–159.

Yasuda, Y. (2004) Monsoons and religions. In *Monsoon and Civilization*, ed. Y. Yasuda and V. Shinde, New Delhi: Lustre Press, pp. 319–338.

Yasunari, T. (1980) A quasi-stationary appearance of 30 to 40 day period in the cloudiness fluctuations during the summer monsoon over India. *J. Meteor. Soc. Japan*, **58**, 225–229.

Yasunari, T. (1981) Structure of an Indian Summer monsoon system with around 40-day period. *J. Meteor. Soc. Japan*, **59**, 336–354.

Yin, A., Kapp, P. A., Murphy, M. A. *et al.* (1999) Significant late Neogene east–west extension in northern Tibet. *Geology*, **27**, 787–790.

Yoshikawa, S. and Inouchi, Y. (1991) Tephrostratigraphy of the Takashima-oki boring core samples from Lake Biwa, central Japan. *Earth Sci.* (*Chikyu Kagaku*), **45**, 81–100.

Yoshikawa, S. and Inouchi, Y. (1993) Middle Pleistocene to Holocene explosive volcanism revealed by ashes of the Takashima-oki core samples from Lake Biwa, central Japan. *Earth Sci.* (*Chikyu Kagaku*), **47**, 97–109.

Young, T. C. (1995) The Uruk World System: the dynamics of expansion of Early Mesopotamian civilization. *Bull. Amer. Sch. Orient. Res.*, **297**, 84–85.

Zachos, J. Pagani, M., Sloan, L., Thomas, E. and Billups, K. (2001) Trends, rhythms and aberrations in global climate 65 Ma to Present. *Science*, **292**, 686–693.

Zahn, R. (2003) Monsoon linkages. *Nature*, **421**, 324–325.

Zeitler, P. K., Koons, P. O., Bishop, M. P. *et al.* (2001) Crustal reworking at Nanga Parbat, Pakistan; metamorphic consequences of thermal-mechanical coupling facilitated by erosion. *Tectonics*, **20**, 712–728.

Zhang, P., Molnar, P. and Downs, W. R. (2001) Increased sedimentation rates and grain sizes 2–4 Myr ago due to the influence of climate change on erosion rates. *Nature*, **410**, 891–897.

Zhao, J., Wang, Y., Collerson, K. D. and Gagan, M. K. (2003) Speleothem U-series dating of semi-synchronous climate oscillations during the last deglaciation. *Earth Planet. Sci. Lett.*, **216**, 155–161.

Zhao, Q., Wang, P., Cheng, X. *et al.* (2001a) A record of Miocene carbon excursion in the South China Sea. *Sci. China, Ser. D*, **44**, 943–951.

Zhao, Q., Wang, J., Cheng, X. *et al.* (2001b) Neogene oxygen isotopic stratigraphy, ODP Site 1148, northern South China Sea. *Sci. China, Ser. D*, **44**, 934–942.

Zhao, W. L. and Morgan, W. J. P. (1987) Injection of Indian crust into Tibetan lower crust; a two-dimensional finite element model study. *Tectonics*, **6**, 489–504.

Zheng, H., Powell, C. M., An, Z., Zhou, J. and Dong, G. (2000) Pliocene uplift of the northern Tibetan Plateau. *Geology*, **28**, 715–718.

Zickfeld, K., Knopf, B., Petoukhov, V. and Schellnhuber, H. J. (2005) Is the Indian summer monsoon stable against global change? *Geophys. Res. Lett.*, **32**, L15707, doi: 10.1029/2005GL022771.

Zielinski, G. A., Mayewski, P. A., Meeker, L. D. *et al.* (1996) Potential atmospheric impact of the Toba mega-eruption approximately 71 000 years ago. *Geophys. Res. Lett.*, **23** (8), 837–840.

Further reading

Balsam, W., Ji, J. and Chen, J. (2004) Climatic interpretation of the Luochuan and Lingtai loess sections, China, based on changing iron oxide mineralogy and magnetic susceptibility. *Earth Planet. Sci. Lett.*, **223**, 335–348.

Balsam, W., Ellwood, B. and Ji, J. (2005) Direct correlation of the marine oxygen isotope record with the Chinese Loess Plateau iron oxide and magnetic susceptibility records. *Palaeogeog., Palaeoclim., Palaeoeco.*, **221**, 141–152.

Banakar, V. K., Galy, A., Sukumaran, N. P., Parthiban, G. and Volvaiker, A. Y. (2003) Himalayan sedimentary pulses recorded by silicate detritus within a ferromanganese crust from the central Indian Ocean. *Earth Planet. Sci. Lett.*, **205**, 337–348.

Chakraborty, S. and Ramesh, R. (1993) Monsoon-induced sea surface temperature changes recorded in Indian corals. *Terra Nova*, **5**, 545–551.

Charles, C. D., Hunter, D. E. and Fairbanks, R. G. (1997) Interaction between the ENSO and the Asian monsoon in a coral record of tropical climate. *Science*, **277**, 925–928.

Chen, L. X. (1991) *East Asia Monsoon*. Beijing: Meteorology Press of China, pp. 1–262 (In Chinese).

Clift, P. D. and Gaedicke, C. (2002) Accelerated mass flux to the Arabian Sea during the Middle-Late Miocene. *Geology*, **30**, 207–210.

Davis, M. E., Thompson, L. G., Yao, T. and Wang, N. (2005) Forcing of the Asian monsoon on the Tibetan Plateau; evidence from high-resolution ice core and tropical coral records. *J. Geophys. Res.*, **110**, D4.

Fluteau, F. (2003) Earth dynamics and climate changes. *Comp. Rend., Acad. Sci., Geosci.*, **335**, 157–174.

Glennie, K. W., Singhvi, A. K., Lancaster, N. and Teller, J. T. (2002) Quaternary climatic changes over southern Arabia and the Thar Desert, India. In: The *Tectonic and Climatic Evolution of the Arabian Sea region*, ed. P. D. Clift, D. Kroon, C. Gaedicke and J. Craig. *Geol. Soc., Lond., Spec. Publ.*, **195**, 301–316.

Goodbred, S. L. (2003) Response of the Ganges dispersal system to climate change; a source-to-sink view since the last interstade. *Sed. Geol.*, **162**, 83–104.

Ji, J., Chen, J., Balsam, W. *et al.* (2004) High resolution hematite/goethite records from Chinese loess sequences for the last glacial–interglacial cycle; rapid climatic response of the east Asian monsoon to the tropical Pacific. *Geophys. Res. Lett.*, **31**, 3.

Li, B., Jian, Z., Li, Q., Tian, J. and Wang, P. (2005) Paleoceanography of the South China Sea since the middle Miocene; evidence from planktonic foraminifera. *Mar. Micropaleo.*, **54**, 49–62.

Liu, L., Chen, J., Ji, J. and Chen, Y. (2004a) Comparison of paleoclimatic change from Zr/Rb ratios in Chinese loess with marine isotope records over the 2.6–1.2 Ma B. P. interval. *Geophys. Res. Lett.*, **31**, L15204, doi: 10.1029/2004GL019693.

Loschnigg, J. and Webster, P. J. (2000) A coupled ocean–atmosphere system of SST regulation for the Indian Ocean. *J. Clim.*, **13**, 3342–3360.

Lu, H., Zhang, F. and Liu, X. (2003) Patterns and frequencies of the East Asian winter monsoon variations during the past million years revealed by wavelet and spectral analyses. *Global Planet. Change*, **35**, 67–74.

Ma, Y., Fang, X., Li, J., Wu, F. and Zhang, J. (2004) Vegetational and environmental changes during late Tertiary–early Quaternary in Jiuxi Basin. *Sci. China, Ser. D, Earth Sci.*, **34**, 107–116.

Naidu, P. D. and Malmgren, B. A. (1995) A 2 200 years periodicity in the Asian monsoon system. *Geophys. Res. Lett.*, **22**, 2361–2364.

Qiao, Q. M. and Zhang, Y. G. (1994) *Climatology of Tibet Plateau*. Beijing: Meteorology Press of China, pp. 1–250.

Ruddiman, W. F. and Kutzbach, J. E. (1990) Late Cenozoic uplift and climate change. *Trans. R. Soc. Edin. Earth Sci.*, **81**, 301–314.

Savidge, G., Elliott, A. E. and Hubbard, L. (1992) The monsoon induced upwelling system of the southern coast of Oman. *Mar. Geol.*, **104**, 290–291.

Schulz, H., von Rad, U. and Ittekkot, V. (2002) Planktic foraminifera, particle flux and oceanic productivity off Pakistan, NE Arabian Sea; modern analogues and application to the palaeoclimatic record. In: *The Tectonic and Climatic Evolution of the Arabian Sea Region*, ed. P. D. Clift, D. Kroon, C. Gaedicke and J. Craig. Geol. Soc., Lond., Spec. Publ., 195, 499–516.

Tian, J., Wang, P., Cheng, X. and Li, Q. (2002) Astronomically tuned Plio-Pleistocene benthic $\delta^{18}O$ record from South China Sea and Atlantic-Pacific comparison. *Earth Planet. Sci. Lett.*, **203**, 1015–1029.

Tudhope, A. W., Lea, D. W., Shimmield, G. B., Chilcott, C. P. and Head, S. (1996) Monsoon climate and Arabian Sea coastal upwelling recorded in massive corals from southern Oman. *Palaios*, **11**, 347–361.

Van Campo, E. (1986) Monsoon fluctuations in two 20,000-yr B. P. oxygen-isotope/pollen records off southwest India. *Quat. Res.*, **26**, 376–388.

Wang, P., Prell, W., Blum, P., *et al.* (2001b) *Proc. ODP, Init. Repts.*, **184**, College Station, TX: Ocean Drilling Program.

Webster, P. J. and Chou, L. C. (1980) Seasonal structure of a simple monsoon system. *J. Atmos. Sci.*, **37**, 354–367.

Yuan, D., Cheng, H., Edwards, R. L. *et al.* (2004) Timing, duration, and transitions of the last interglacial Asian monsoon. *Science*, **304**, 575–578.

Zheng, H., Powell, C. M., Rea, D. K., Wang, J. and Wang, P. (2004) Late Miocene and mid-Pliocene enhancement of the east Asian monsoon as viewed from the land and sea. *Global Planet. Change*, **41**, 147–155.

Index

Printed in the United States
By Bookmasters